A Series of Essays and Research Papers

Greenhouse Warming and Nuclear Hazards

A Series of Essays and Research Papers

Greenhouse Warming and Nuclear Hazards

Peter Fong

Emory University, USA

World Scientific

NEW JERSEY · LONDON · SINGAPORE · BEIJING · SHANGHAI · HONG KONG · TAIPEI · CHENNAI

Published by

World Scientific Publishing Co. Pte. Ltd.

5 Toh Tuck Link, Singapore 596224

USA office: 27 Warren Street, Suite 401-402, Hackensack, NJ 07601

UK office: 57 Shelton Street, Covent Garden, London WC2H 9HE

British Library Cataloguing-in-Publication Data
A catalogue record for this book is available from the British Library.

GREENHOUSE WARMING AND NUCLEAR HAZARDS
A Series of Essays and Research Papers

ISBN-13 978-981-256-422-1 (pbk)
ISBN-10 981-256-422-5 (pbk)

To K. L. Cheng

For his foresight to the age of wisdom away from the age of foolishness and his first proposal for a device to use nuclear waste for the improvement of health*.

K. L. Cheng, *Microchemical Journal* **72**, 113–114 (2002).

It was the best of times, it was the worst of times, it was the age of wisdom, it was the age of foolishness, it was the epoch of belief, it was the epoch of incredulity, it was the season of Light, it was the season of Darkness, it was the spring of hope, it was the winter of despair, we had everything before us, we had nothing before us, we were all going direct to Heaven, we were all going direct the other way—in short, the period was so far like the present period, that some of its noisiest authorities insisted on its being received, for good or for evil, in the superlative degree of comparison only.

A TALE OF TWO CITIES BY *Charles Dickens*

"Chiü Nüan was exiled, wandering on the river bank. The fisherman spotted him and asked: "Aren't you the Alderman of the Third Precinct? Why are you so anguished?" Chiü Nüan answered: "The entire world is muddy but I only am clean; everybody is drunk but I only am sober. That's why I am exiled." The fisherman then asked: "If the whole world is muddy, why not dig up the dirt and stir up the waves? If everybody is drunk, why not chew up the mash and suck up the alcohol?" The reply: "I understand the newly bathed cleans the robe and the newly washed dresses the hair. How could I subject my pristine body to the filthy whirlpool. I would rather go to the sparkling Miro River and live with the fishes and shrimps."...Lamented the fisherman: "Oh! Clean water of the turbulent river! Wash my tassels! Muddy water of the turbulent river! Wash my feet!" Then left without further words.

The Fisherman, by Chiü Nüan* 屈 原, ca. 500 B.C.

* More on pp. 26, 275-294.

CONTENTS

Foreword

There are three major global environmental problems: greenhouse warming, nuclear hazards and the ozone hole. The ozone problem has been solved (Nobel Prize awarded). This book deals with the remaining two issues.

The current greenhouse *warming* theory would be correct on Earth without the ice ages, such as in the Cretaceous Period (Dinosaurs). Then we can study the atmosphere based on *stable equilibrium*, by which the increase in carbon dioxide by fossil fuels will lead to global warming. However, our Earth today is still in an ice age (in the *interglacial* of *Pleistocene*), in which the polar ice caps and the cloud layers act as *thermostats* to stabilize temperature and to avoid global warming. Ice age is in *neutral equilibrium*, due to the *phase transitions* of water in which the temperature is unchanged, leading to *thermostat* effects.

To be specific, upon adding greenhouse *heat* due to carbon dioxide, Earth will respond by increasing evaporation of ocean water and *cloud coverage*. The cloud so increased will cool down Earth by reflecting away more sunlight. This leads to the *crucial* result of dissipating the greenhouse heat *entirely* so that there will be no increase in *temperature* to cause global warming. These conclusions are verified by the observed data of cloud, precipitation and temperature changes in the past century and calculations of cloud dissipation based on the ERBE satellite data. The *exact cancellation* of global warming is not by accident. This is how the invisible hands of thermodynamics work through the ice age *thermostats*. The study of global warming is thus closely related to the origin of ice ages.

The origin of ice ages has been a century-old, top-mystery problem, which I started to examine in 1982, ten years before the IPCC study of greenhouse warming in 1990. I solved the ice age origin problem, from which the exact cancellation of greenhouse heat comes as a matter of course, not as a mathematical freak. However official theoretical studies based on *stable equilibrium* led to widely different warming from 0°C to 5°C. Scientists want a $1 billion super-computer to resolve the differences in ten years.

Experimentally no warming was found and the problem has always been the *missing greenhouse heat*. The major advocate IPCC, after five years, has found *shimmers* of greenhouse heat. After ten years (2001) it concluded that there is no warming now due to the cooling effect of the polluted aerosols (actually our clouds) but warming will return when future pollution cleaning would have removed the aerosols.

Indeed warming would begin when the increased cloud would have covered the entire Earth (as in Cretaceous). That would take 1200 years in the current ice age Earth. But fossil fuels would be exhausted much earlier, in 300 years. Thus the greenhouse warming is a *moot issue*; ironically it is due to too little fossil fuel to bring it about. The bitter dispute ends in a bizarre happy finale, with no sour losers, no crisis and no scandal, that deserves a grand celebration.

The nuclear issue is the hazard of low-level radiation. By the official doctrine of *linear extrapolation without threshold*, it is hazardous all the way to zero radiation. Although the amount is small the total added up could be large. It makes nuclear waste disposal a serious problem. Sporadic evidence has generated a new view that there is a *threshold* below which it becomes *beneficial*. Then waste disposal will not be serious.

To resolve the low-dose issue we studied the extremely voluminous death statistics of leading diseases—heart, cancer, stroke, lung, diabetes—of the entire population of the 50 states of US separately over the past 50 years. Then we correlated them with the low radiation exposure of the 50 states—*natural* due to altitude differences and *artificial* due to radiation from Nevada nuclear tests at different distances away.

We found that eight high mountain states with high radiation have their cancer and other death rates reduced by 25%. The Nevada nuclear tests have reduced Nevada's cancer death rate of 25% after five years of tests. Other states have reduced rates diminishing with distance from Nevada and to zero for Hawaii and Alaska. Since this study has used up all available data, leaving no room for any other studies, the results must be the ultimate truth that cannot be challenged. The beneficial effect of low-level radiation is thus established without doubt and quantitatively in exact proportion.

Since all major diseases benefit from low level radiation, we can expect a correlation of *longevity* with high radiation. In fact, in the Kerala State of India, where the radiation level is 20 times higher than normal (by Th mine), the life expectancy is 10.5 years longer than India as a whole.

The unavoidable question is why low level radiation can render health benefits against common sense. There is a reasonable explanation. Since Earth was formed, it has been hit by background radiation, which is hazardous to life. Life then fought back by developing a primitive immune system, just to produce *an anti-oxidant* molecule to remove the *free radicals* generated by radiation that harms life. Later the system can be stimulated by low-level radiation to generate more anti-oxidants to prolong life. A major hazard against nuclear power is removed. Even nuclear trash can be turned into treasure to benefit mankind.

The book thus removes two pernicious stumbling blocks against energy development and should help the economic expansion of the world.

P. Fong

Preface

This book is to communicate the new research results on two major public issues—greenhouse warming and nuclear hazards—to the interested and affected citizens, who, without specialized technical knowledge, care and follow the issues for their own and for the nation. The research is carried out according to the august regimen of theoretical and experimental physics, seared in crucible, tempered by contention. The results are iron-clad against the prevailing views. The advocates of greenhouse warming and nuclear hazards must confront the challenge—to rebut or to quit, lest they be classified as fraudulent and dishonest.

     *   *   *   *   *   *   *   *   *   *   *   *

The issue of global *warming* can be settled by three smoking-gun evidences experimentally (see the first paper hereafter). They show that the greenhouse *heat* is entirely dissipated to outer space by the cloud generated by the greenhouse *effect*. Therefore there is no greenhouse heat left on earth to generate global *warming*. That is the end of the line. For practical purposes any further discussion is needless. The commander-in-chief is ready to send the order of attack without hesitation. On the other hand from the *theoretical* point of view much is needed to explain and establish the scientific basis of the experimental truth and to explicate the misleading ideas that are prevalent.

All basic principles of science on the problem are well known. What is missing is a branch of science that is dedicated to a specific category of phenomena covering the problems concerned, such as elastic body dynamics for elasticity problems. Both global warming and the ice age are climate problems of the *Pleistocene* epoch, which are characterized by two global *thermostats*: one is the permanent ice in the polar regions, the other is the clouds in the sky. By ice-water, and water-vapor *phase equilibrium* respectively they tend to maintain a constant temperature in a *neutral equilibrium*, and thus get rid of all multifarious and complicated intruding influences, making possible to reach a crisp, simple conclusion. This is the way to reach the conclusion of no global warming.

This overriding fact determines all climate problems in Pleistocene, which can be treated likewise. Therefore a new branch of science *Pleistocene climatology* is called for, analogous to *dynamics of elastic bodies* to deal with a special category of phenomena in a clearly defined domain of scientific study.

The *canonical formulation* of a branch of science of mechanics is to set up an *equation of motion* which consists of 3 parts: 1, the *universal laws* of Newton, 2, the *branch laws* of the *specific branch*, such as Hooke's law for the elasticity branch, and the neutral equilibrium law in Pleistocene climatology. 3, the *specific laws of force* for the multifarious specific problems within the branch, such as the various forms of the potential term $V(\underline{r})$ in the Schroedinger equation. The specific forcings for the greenhouse and the ice age problems are the *greenhouse heat forcing* and the *Milankovitch forcing*. These two problems have been solved independently in history. All other problems in Pleistocene are traceable to a specific forcing, for example, the *Heinrich events* may be traced to comet impacts. While the ice age is not our problem, it was brought up for relevance.

In *Pleistocene climatology* all climate problems in the ice age epoch can be treated *systematically* (not *haphazardly*) like elastic problems in dynamics of elastic bodies from a general equation of motion of this branch and specific forcings for specific problems under consideration. All climate problems of human concern are limited to the last few millennia of this epoch. For practical purposes we need to search no further for climate studies.

For the climate problems there is a *new* 4th part of the equation of motion in addition to the above three, the *feedback effects*, such as the *albedo* and *infrared* feedbacks. These are forces not previously existent in the system but induced by the phenomenon once it is underway and must be included in the

Newton's second law to solve the problem. This complicates the matter. Worse still, the feedbacks, as in ice ages, could be two orders of magnitude greater than the original forcing, which completely mess   up the strategic outlook of the problem. Even more traumatic is that the feedbacks are so numerous and mutually interacting, like the contestants entering the Thirty Years War, that they completely messed up the war plan and ended up in complete ruin of all participants in the Deutschereich. As a matter of fact, for our interest, it is the *cloud feedback* that gets rid of the greenhouse *heat*, suggesting no *warming*. But that is not yet the end of the problem because all the other feedbacks have not been accounted for.

The past performances of the greenhouse and ice age studies are modern examples of such a disaster. The greenhouse warming problem involves dozens of feedbacks, only a few have been studied. A few new ones raised by Richard Lindzen are enough to cast doubt on the current conclusion. This line of work now stands still awaiting the arrival of super-computer ten years hence to save the impotence of the current generation of computers to tackle the complex problem.

In *Pleistocene climatology* most feedbacks are overwhelmed by the equilibrium condition already contained in part 2 and need not be considered separately again, greatly simplifying the problem. The part 4 on feedbacks can be treated neatly and conclusively. For our interest, it can be shown that only the cloud feedback needs to be considered and all others have no part to play. That solves the problem. No super-computer is needed. Save $1 billion.

Just as in the canonical theory of mechanics, *canonical generalized variables* may be introduced which can be solved more readily. In the greenhouse problem the canonical variables are the *mass of precipitation increase* m_1, the *mass of the cloud increase* m_2, and the *global temperature increase* T. They can be solved from the *canonical equations*. m_1 and m_2 are sizable quantities and can be compared with experiments. The results of comparison are perfect, proving the theory and the reality of the greenhouse *effects*. The result of T happens to be *zero* up to 3000 years hence, which is the limit of the theory. Empirical result is *zero now* and whatever happens in the future is relevant only for the next 300 years, after which fossil fuels will be completely depleted and the issue becomes moot. In the next 300 years, which is a blink in the Pleistocene time scale, only a charlatan can make a mountain out of a molehill (pp.79, 100). Thus any and all discussions of greenhouse warming are moot. The Kyoto Protocol is thus as realistic as Ponce de Leon's search for the Fountain of Youth.

That the canonical variable T is a constant is not at all unusual. The *angular momentum p*, as a cononical variable, is a constant. Like other *constants of the motion*, the conservation of angular momentum implies a "*symmetry*" property of the space with respect to *rotation*. The constancy of T as a canonical variable, also implies a global "*symmetry*" property of the climate system, that is, the neutral equilibrium condition. In the ice age case the Melankovitch forcing acts as "*symmetry breaking*,"—a notion fundamental in high energy physics.

Thus Pleistocene climatology has reached the ultimate canonical destination, leaving all the roundabout, grotesque detours to the wasted history.

    *   *   *   *   *   *   *   *   *   *   *   *

With the greenhouse warming issue out of the way, a canonical *energy policy* emerges naturally thus: (1), *First,* fossil fuels should be banned for energy production use, not because it generates carbon dioxide and greenhouse warming but because of the double reasons as follow: (a), It generates air and water pollution, which are our foremost concern. (b) It depletes unrenewable natural resources of *organic carbon*, which, in the form of plastics, is the main source of supply of necessary material to sustain modern civilization after most mineral resources, including all metals, will be depleted in a century. To burn the fossil fuels for energy generation is as horrible as burning a copy of Gutenberg Bible to warm a cup of coffee.

(2), *Second*, without the fossil fuel energy supply, what will be the main energy source for the future? Hydraulic, wind, geothermal, bio-mass, tide, deep ocean energy sources have all been explored. Most of them can supply only about 1% of the energy needs except the hydraulic, which can supply a little more, a few percent. Thus the bulk of the energy needed must be supplied from the only energy resource remaining—the *nuclear power*, which has already supplied about 10% of all energy consumed (20% of all electricity generated). [For details see my book on energy and environment by *Macmillan*.] This leads to an anathema to the general public that is indoctrinated to the notion of nuclear catastrophe. The second part of this book is dedicated to alleviate this fear.

It must be stated first of all that the objective of an energy policy is to guarantee the full development, economic and social, of the nation, not to pursue specific objectives, such as "the nuclear free homestead" for its own sake. Since pollution is a major social issue, it ranks as a top priority in energy considerations. In our energy advocacy environmental preservation is always an overriding concern.

*　　*　　*　　*　　*　　*　　*　　*　　*　　*　　*　　*

Before addressing the nuclear hazards it is to be noted that the greenhouse warming and nuclear hazards are two major issues in modern times, involving environmental, economic, social, political and spiritual aspects. Both in all aspects have been discussed in the public with myriads of unresolvable issues. Those aspects that are well known and uncontroversial will not be repeated here. What will be done here, as in the greenhouse problem, is to strike at the heart of the scientific issue. The rest will follow naturally. The heart of the nuclear issue is the *health effects of low level nuclear radiation*.

Low level radiation is defined as radiations of a strength comparable to *natural background radiation,* which is omnipresent, created by God before humans came to the earth, and thus no one can question its *raison d'etre,* and thus is a natural standard to use for comparison. Its strength is *100 millirem per year* at the sea level, which every body is exposed to. It is a natural standard for comparison with other sources of concern, such as dental X-rays, nuclear plant emissions, etc.

Besides the normal background radiation, every body is exposed to higher background radiation from high altitude (Denver has a level twice larger than at sea level), from medical diagnosis and treatment, from all sorts of nuclear technologies (including wrist watches and smoke alarms), from nuclear power plants (in normal operation and in accidents) and from nuclear bombs (in tests and in war). No one can escape these and it behooves everybody to learn the scientific facts affecting human health which will be summarized in the following.

The conclusions obtained are based on *vital statistics* of the United States populations of the 50 states in the past 50 years, which are contained in a truck load of government publications that is available in any good-sized library. This data set is enough to settle all important questions conclusively. Nothing more is needed (unless U.S. have annexed 50 more States and have lived 50 more years to necessitate an updating of a few percent). No further experiments are needed, necessary, desirable, and possible. No question on the veracity of the scientific truth is justified. This is it. No one had, have, can have, and will have different conclusions from what is obtained from this data set described in the following.

However, this job has not been done by the establishment, by the energy industries for their self interest, by the academe for scholastic studies, and by the counter-culture for their Quixotic and rebellious pursuits. It does not need new august laws as Newton's and Hooke's, and new epoch making formulations as in the greenhouse problem. It just needs hard spade work to go through the dry as saw dust statistics to sift out the missing diamonds. It does not need super-genius; it just needs super-handyman. Without such a knowledge there is no way to set technical standards to guarantee health and safety while maximizing efficiency and benefits. Failure to do that is irresponsibly delinquent at best and criminal negligent at worst. What actually happened was that nothing has been done. No truths emerged. Illusions overflow. Fallacies abound. This is a *mad, mad, mad World*, as in the movie of that title.

["Everybody is drunk and I only am sober. But I will not chew up the mash and suck up the alcohol." said *Chiü Nüan* before committing suicide 2500 years ago. This is the same for the global warming issue and the Kyoto Protocol.]

The conclusions of the nuclear hazards issues we obtained are: (1). A doubling of the background radiation (such as moving to live in Denver) has the effect of *reducing cancer death rate by 25%*. (2) Similar results are obtained for *heart diseases, stroke, and most of the leading diseases* causing death. (3) As a result low level radiation has an effect to *extend life span*. In fact, in Kerala, India, *life expectancy is increased 10.7 years* due to an increase of background radiation of 20-fold due to the thorium mine deposits there. (4) The reduction of cancer and the increase of radiation level of the 8 mountain states of the Rocky Mountain region as functions of the altitudes of the 8 states *follow the same exponential curve*, thus establishing the *causal* relation of the radiation and cancer reduction beyond *circumstantial correlation*. (5) Cancer study of Rocky mountain region agrees with longevity study of Kerala quantitatively. The above results are related to *natural* low level radiation.

(6) *Artificial* low level *radiation from nuclear bomb tests* in air in the 1950s in Nevada, instead of increasing cancer as first expected and feared, has actually *reduced cancer deaths in the U. S. by half a million.* (7) *Nevada* cancer

death rate was *reduced by 25% in 4 years* after tests began and continued for a decade. (8) *Other states* have less cancer reduction, the farther the lesser. (9) The farthest states *Alaska and Hawaii* have no reduction. Comparing these iron-clad conclusions with the prevailing sentiment against nuclear power, you now can realize what a mad, mad, mad world we are in.

(10) The quantitative law relating *radiation and cancer death reduction* is the same in bomb radiation (created by humans) as in natural radiation (created by God). Thus bomb radiation is not particularly *evil* as many had in mind. A repeat of Newton's work in showing that the laws of mechanics are the same in Heaven as on earth, thus ushering in the modern world and leaving medieval age behind. Now the same with the medieval view of nuclear hazards.

The 50 States provide 50 independent observers monitoring the same experiment for 50 years on both natural and bomb radiations. No experiments can be more comprehensive. And it did not cost a penny of research funds (which would run into millions of dollars in comparable projects) from the government and others except my research time and the work of my student David Bovi whose color charts for this work enhance the appearance of the book. His calculation of the *population weighted average altitudes* of the 8 mountain states, as shown in 4 digit figures in the color charts, is an indication of the extent of spade work that has gone into the project.

It is ironic that the *Airborne Nuclear Test Ban Treaty* was promulgated expeditiously and put into effect in the middle 1950s for humanitarian reasons. In retrospect it was an re-enactment of the *Chicken Little* story of sky-is-falling. The fallen sky is actually an apple and is good for eating. Nuclear tests after the Test Ban Treaty had gone underground and 1500 bombs were exploded. If exploded in air it would have saved 1.5 million American lives from cancer deaths. And increased life span for every Americans except Alaska and Hawaii.

Suggested high school science project: How many millions more lives would be saved from heart diseases and strokes due to the bomb tests? Use government published statistic cited and the information in paper (C) later. Establishment! Vacate the driver's seat for the high school project makers.

It is worth mentioning that the Conference *A Decade after Chernobyl* sponsored by United Nations, European Union, World Health Organization, Russia, White-Russia and Ukraine has declared that there was no increased cancer deaths after the Chernobyl accident. Ten years before, right after Chernobyl, thousands of European women have aborted their unborn children for fear of giving birth to deformed babies. An evil doctrine of a cult has left thousands of murders of the unborn in their trail. This is truly a mad, mad, mad world.

While the cancer fear is declining, one last obstacle still remains against the acceptance of nuclear power—the issue of *nuclear waste disposal*. Nuclear radiation is indestructible. Once created it lasts forever. It pollutes the pristine earth for eternity and cannot be accepted for humanitarian and spiritual reasons, so the argument goes. However, once we learn, this time with respect to the *nuclear waste*, that its low level radiation is not only *not harmful* but also *beneficial* to health and can *prolong life*, then it is no longer a problem to be feared. In fact it will be a billion dollar gold mine to develop health improving equipment. By folklore, people in Brazil and Chekoslovakia have already taking advantage of the nuclear radiation available from their beach sand and uranium mines for health improving uses. Truly it is from *trash to treasure*.

The basic principle controlling the nuclear health effect is simple. The background radiation is the first *antigen* on earth before life has arrived. Therefore the first *immune system* the evolution process endowed upon life is to generate resistance to radiation, mainly by producing *anti-oxidants* to counter the *free radicals* produced by radiation. It so happens that free radicals are also the natural products of metabolism and are the aggravating factors of most diseases and the main cause of aging due to the decline of the innate ability to cleanse the body of the waste products through aging. Therefore, happily, *radiation immune reaction* happens to be the *right antidote against most disease and aging*. An extra amount of low level radiation by whatever sources will stimulate the immune system to generate more anti-oxidants to relieve suffering and prolong life.

A few hour's exposure to the radiation from the Brazilian beach sand or the Chekoslovakia or Colorado uranium mine tunnel would act as a vacuum cleaner to sweep away the metabolic waste products and make one *feel fresher and younger* after the experience. Thus the origin of the nuclear folklore.

The low level radiation can come from any source, even from the atomic bomb exploded in Hiroshima at a large distance from ground zero. Such a result has been reported by Japanese scientists, and published in *Time*, at a conference in New York on the occasion of the 50th anniversary of Hiroshima and the birth of nuclear medicine that people in such a condition have a longer life span. Similar results are found among the *radium watch dial painters*, most of them caught lung cancer because of exposure to *strong radiation*, but a small group with *low radiation* exposure achieved longevity. This radiation effect on longevity is truly a special Grace from God beyond expectation, a dream pursued by emperors, kings, alchemists without avail, now realized serendipitously. But the age of foolishness took it as a scourge created by humans.

During the height of nuclear hysteria after the major nuclear plant accidents it was fashionable to speak of *nuclear free homestead*. Anything nuclear is untouchable. It is often forgotten that the human body is a nuclear radiation emitter due to the radioactive potassium-40 in the body. In Los Alamos National Laboratory there is a machine, which one can walk in to have his body's radiation strength measured. It usually accounts for one-third of the natural background radiation. In a nuclear free homestead such human body is untouchable. The advocates will have to commit self-immolation to support their argument.

Thus it was an age of wisdom; it was an age of foolishness. We were going straight to Heaven; we were going straight to Hell. So said Charles Dickens more than a century ago. And so it remains to this today. This applies not only to the energy issues of greenhouse warming and nuclear hazards but also to many prevailing controversies in politics, economy, society and spiritual life. All of us should strive to turn the age of foolishness to the age of wisdom.

To begin with energy. Plentiful and clean energy is the life line of civilization beyond the cave culture. To achieve this goal two innovations stand out above all as high priority. First, to convert fossil fuels from energy resources to organic carbon resources for industrial use in the next century. Second, to re-build the American nuclear power industry as substitute of fossil fuels, which is depleting rapidly and requires replacement anyway. Revive the 100 aborted nuclear plants abandoned two decades ago. Go back to *nuclear industry standards* of the 1950s. Then, a nuclear plant, such as the one outside Rochester, New York, was built in 3 years (the same as in Japan now). Today in U.S. it takes 13 years to build. The mere increase of interest costs is enough to double the investment cost and thus double the cost of electricity generated. The increased time and costs are for environmental wrangling and stricter regulations to control the nuclear genie (forgetting it to be the nuclear angel). It is truely a mad, mad, mad world. [Others have been discussed in my book by *Macmillan*.]

The double chastity belts of global warming and nuclear hazards must be liberated, lest Humpty Dumpty had a fall—all the establishment, all the vested interest, all the academe and all the counter-culture could not put Humpty Dumpty back again.

Addendum: A Challenge to the Establishment

This concerns the "Position Statement on Climate Change and Greenhouse Gases" of *American Geophysical Union* dated Dec. 1998. Its gist is that: (1). The complexity and variability of the climate system limited the progress of the work. (2). Some accomplishments are at hand nevertheless. (3). The problem is far from reaching a definitive solution. (4). Sustained efforts should continue.

The current climate study is in a stage of the time of the 19th century after Newton but before Euler, Laplace and Boltzmann. Newton's success of few particle mechanics has been applied and succeeded in the study of the *weather* and our accomplishment is displayed in good prediction of weather for a few days.

Once beyond a few days for a longer time we deal with *climate* and are faced with the problem of complexity and variability due to a large number of degrees of freedom. Following the success of weather, the attempt is to extend the Newtonian treatment by super-computer but the success has been limited.

Mechanics of a large number of particles are of course complex and unwieldy. If totally orderless then there is no hope, and not a penny should be spend on the research which is destined to be in vain. However, if some regularity appears out of the chaos, then a systematic treatment can be done to simplify the complexity. Thus in the 18th and 19th century Euler and Laplace developed the dynamics of rigid bodies, elastic bodies and fluids, which are extremely fruitful. The work showed that only a few degrees of freedom (6 for rigid body) are significant, all others are side-tracked. The study of the spinning top would have impressed Newton. There is no need to study the dynamics of every particle of the spinning top by Newton's laws to trace out the motion of all individual particles through super-computer. 99.99% of the information so obtained is a useless quagmire. The only significant information can nevertheless be obtained by solving the Euler's equation without computer. Finally came

Boltzmann with the study of equilibrium, in which only the equilibrium condition is the meaningful physical information, all others are obviated.

Climate study has not gone through such a revolution. The reliance on super computer to exhaust the study of all degrees of freedom of the climate is like that of the study of the spinning top by super computer. Yet we just have a breakthrough. In spite of the complexity of climate there is strict regularity as seen in the ice ages, indicating the existence of simplifying principles to establish a new science *Pleistocene climatology* like rigid body dynamics. This time the special condition is the presence of a permanent body of ice (thus Pleistocene), with the specific simplifying principle embodied in the neutral equilibrium condition of phase transition of ice and water. As Boltzmann has done, the equilibrium condition staffs off all trivial degrees of freedom and simplifies the complexity of the problem. The very fact that in the past century no decent theory of ice ages has been developed means that the climate study has not gone beyond 19th century until recently. With the new work the *complexity issue* is resolved.

The *variability issue* can be resolved in another manner, like a jigsaw puzzle, also demonstrated in the ice age problem. There is only one way to piece a jigsaw puzzle together—there can be no other way. Once it is done, that is it.

With the main issues resolved, the climate problem can be solved crisply. Concerning greenhouse warming whatever fragmentary successes obtained so far are meaningless; all yet unsolved points are irrelevant. There is no more holy grail to be pursued. Only a few moping actions remain, such as the study of the merits and demerits of increased precipitation and cloudiness. The old-time *position statement* is out-dated. At one time, position statements on the issue of round-or-flat earth were periodically issued to review the progress, but the last one has concluded the matter long ago. Now we can sail west bravely without having to worry about reaching the end of the world and then dropping into Hell.

Introduction
Greenhouse Warming and Nuclear Hazards

Three smoking guns prove the falsity of greenhouse warming

Nuclear waste disposal, from trash to treasure

Cancer death rates of 50 states of U.S. related to natural and nuclear-bomb-test-induced low level radiation

Three Smoking Guns Prove the Falsity of Global Warming

Peter Fong

Emory University, Atlanta, GA 30322

Smoking gun #1. Observed facts are that the cloud coverage of the earth has increased 4.1% in the past 50 years (Warren et al, see Figure 1 with reference). Clouds are known to reflect away sunlight and thus cool down the earth. The power rate of cooling is 16.6W/m^2 given by the ERBE Satellite. Then every physics freshman can calculate the total power of cooling for a time of doubling carbon dioxide to be 3.98W/m^2 which just compensates off the *theoretical* greenhouse warming power for doubling carbon dioxide 4W/m^2. Thus no greenhouse warming at all for the current climate condition. The 4W/m^2 is the *only* and *real* concern of greenhouse warming. 4.1%/50 yr is the best of observed data available; all others are either trivial or irrelevant. The warming issue is firmly settled. Kyoto Protocol is thus as quaint and moot as taunting for women suffrage.

Smoking gun #2. Observed facts are that precipitation (rainfall) of the earth has increased 7.8% in the past century (Boden et al, Trends '93, ORNL, 1994). See Figures 2a,b,c with detailed references. It can be shown that the latent heat released in increased rainfall is about equal to that absorbed in evaporation of sea water by a greenhouse warming of 2°C by the increase of vapor pressure. This means the greenhouse heat is used only to increase rainfall, not to warm up the earth. The data used are the largest changes in the longest time ever, thus prove no warming strongly.

Smoking gun #3. The rate 4.1% change of 50 years of cloud turns out to be about same as 7.8% change of 100 years for rainfall, proving quantitatively the causal relation of rainfall and cloud, and double checking

3

Warren, S.G., Hahn, C.J., London, J., Chervin, R.M., and R. L. Jenne, 1988. *Global Distribution of Total Cloud Cover and Cloud Type Amounts over the Ocean.* U.S. Department of Energy Publication DOE/ER-0406, Washington DC.

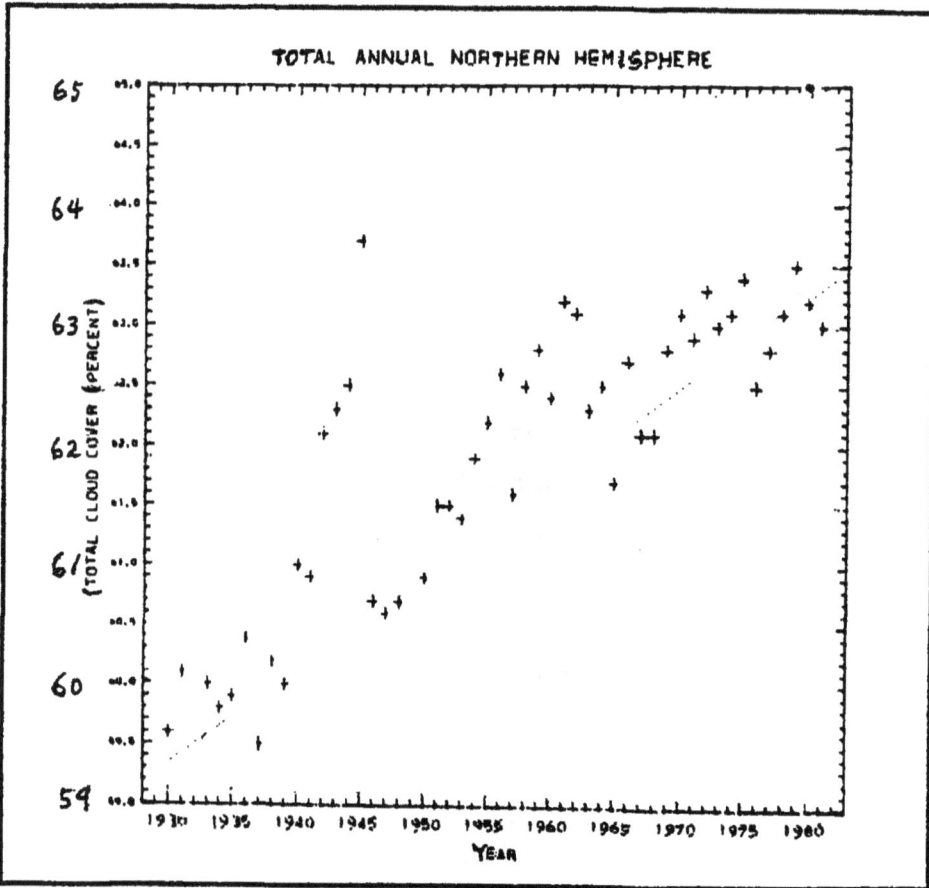

Shipboard observations of Northern Hemisphere cloudiness show a dramatic increase over the years since 1930. When broken down into cloud *types* the data show that the *species highly effective at global cooling (the ocean-surface stratocumulus) shows the most dramatic increase.*

Figure 1

Century-scale series of annual precipitation over the contiguous
United States and Southern Canada

P.Ya. Groisman and D.R. Easterling (continued)

UNITED STATES AND SOUTHERN CANADA
(South of 55° N)

Trends

Groisman and Easterling (1994) performed simple linear regression analyses on the time
series from the entire U.S./Southern Canada study region and the regional North American
time series. They found that the area-average precipitation time series for the entire study
region showed a significant (at the 5% level) upward trend in precipitation amount of +7.6%
per 100 years. They characterize this trend as indicating that during the last century a
large-scale increase in North American precipitation has occurred, owing mainly to
increases in eastern Canada and adjacent regions of the United States (see following sections
detailing regional trends).

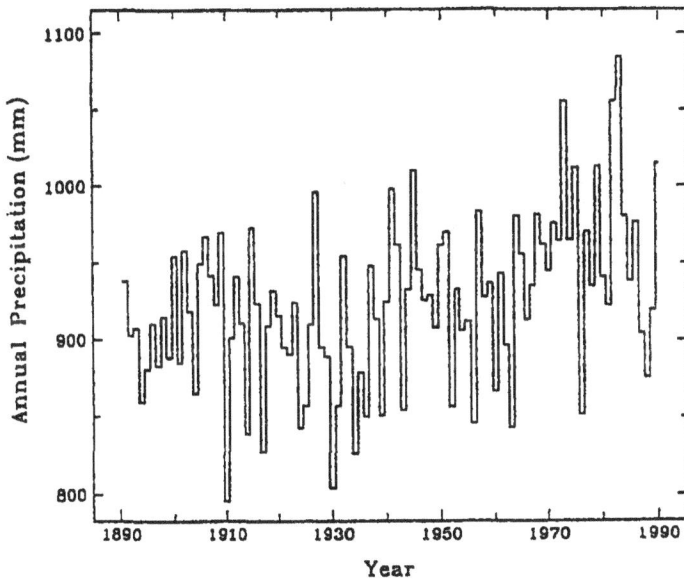

Annual precipitation amount for the United States and Southern Canada.

Figure 2a

area-average precipitation over the main part of the former USSR

P.Ya. Groisman, V.V. Koknaeva, T.A. Belokrylova, and T.R. Karl (continued)

MAIN PART OF THE FORMER USSR
(37°–70° N and 25°–140° E)

Trends

The area-average precipitation time series for the former USSR exhibits a linear increase of 9% per 100 years; this increase is significantly different from zero at the 5% level (Vinnikov et al. 1990). The direction of this trend is consistent with climate-modeling experiments using increased concentrations of greenhouse gases; however, the magnitude of the precipitation trend is considerably greater than the models would suggest (Groisman 1991).

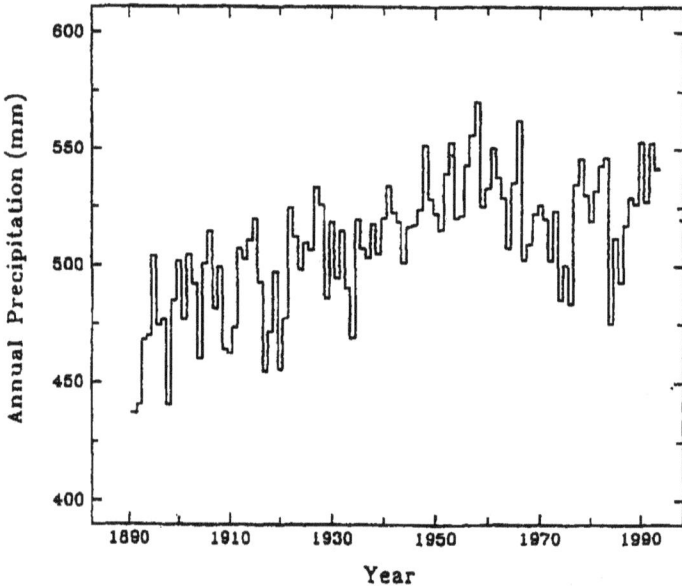

Annual precipitation amount for the main part of the former USSR.

CITE AS: Groisman, P.Ya., V.V. Koknaeva, T.A. Belokrylova, and T.R. Karl. 1994. Area-average precipitation over the main part of the former USSR. pp. 906–910. In T.A. Boden, D.P. Kaiser, R.J. Sepanski, and F.W. Stoss (eds.), *Trends '93: A Compendium of Data on Global Change*. ORNL/CDIAC-65. Carbon Dioxide Information Analysis Center, Oak Ridge National Laboratory, Oak Ridge, Tenn., U.S.A.

Figure 2b

Century-scale series of annual and seasonal precipitation anomaly
for East Asia (15°–60° N; 70°–140° E)

M. Hulme and Z. Zhao (continued)

EAST ASIA
(15°–60° N; 70°–140° E)

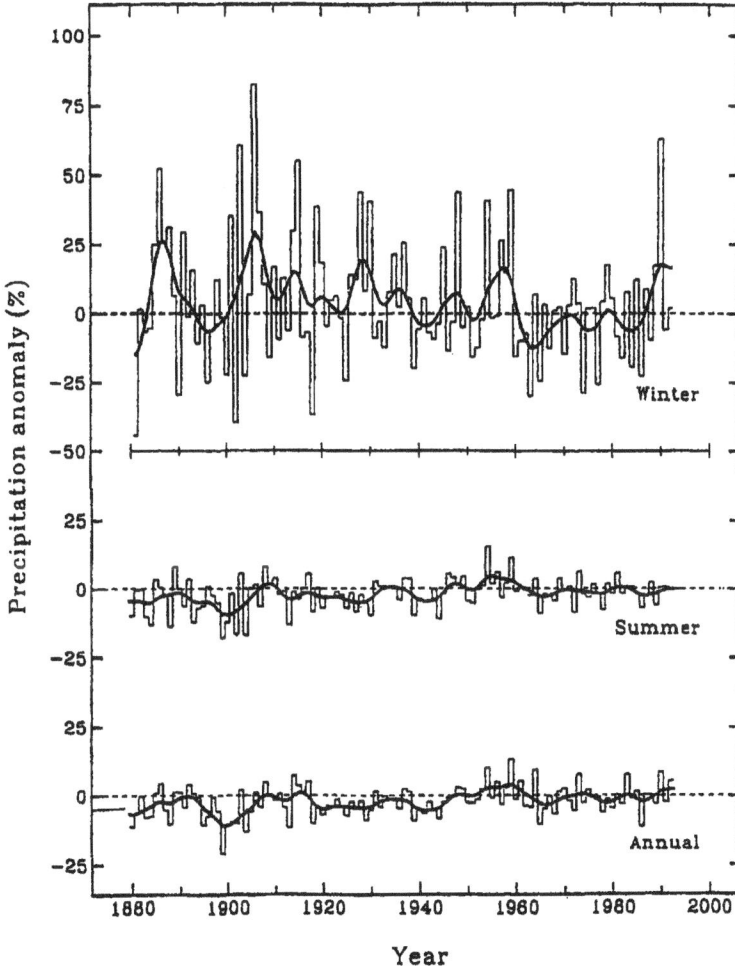

*Annual, winter (December–February), and summer (June–August) precipitation
anomalies for East Asia. Values are percent anomalies from the 1951–80 reference period. The
smooth curves represent the effect of a 10-point Gaussian filter, which suppresses variations
on time scales of less than a decade.*

the accuracy of the two data. This ratio conversion can be done by any child but was missed by all pundits and thus missed the case breaking clue.

These are the *exhibits*. Now the *argument* to establish the *case*. We accept the truth of greenhouse phenomena. We accept the increase of greenhouse gases and the generation of *greenhouse heat*. This heat leads to three *greenhouse effects*; 1, increase of cloudiness, 2, increase of rainfall, 3, possible increase of temperature which is called *greenhouse warming*. The last term means nothing else, which could be *greenhouse effects* such as 1 and 2 but not 3. Most users of the term mistake the use of it with serious misleading consequences. The exhibits show that 1 and 2 (effects) are true and 3 (warming) is false.

The case develops now in a self evident way based on the exhibits: The greenhouse *heat* was first taken up entirely by evaporation of water as latent heat, and recovered later during precipitation at the cloud level. The cloud gets rid of the heat through the cloud's reflection of sunlight. Thus the very greenhouse *heat*, after a few money launderings, is sent back to outer space entirely and exactly without the slightest *warming* of the earth.

Thus there is precisely no greenhouse warming at all. However our rightful concern is global *climate*, not the null global *warming*. The other greenhouse *effects* deserve consideration but have not been done properly. For example, (1), The increase of precipitation is beneficial for increasing the supply of fresh water. It may make oases out of central Asia and Sahara Deserts. However there is an undesirable side effect, the flood. (2), The increase of cloudiness may reduce the UVB radiation reaching the earth surface and may make the ozone hole issue moot but the reduction of insolation may reduce agricultural yield. Thus there should be three suits filed against greenhouse effects 1, 2, 3, separately because they are independent; also the evidences used for prosecution are incommensurate and the remedies offered are mutually irrelevant. In fact suits 1 and 2 have not been filed but are often wrongly dragged into suite 3, resulting in great confusion. This work concerns exclusively on issue 3 *warming*.

There are additional confusion to be avoided. The sea level rise is an *ice age* problem, not related to greenhouse gases. Facts are that polar icesheets are *not* melting and sea level rise does *not* go beyond the secular trend of 2mm/year (more later). If greenhouse heat were used to melt the polar ice, sea level would rise 7 cm/yr, a world apart.

Kyoto Protocol is addressed to global warming with particular attention to fossil fuel and air pollution. As warming is null, it is much ado for nothing. The gavel has struck. There is no room for further discussion. The next section is a *pedestrian* explanation for the scientific basis of the conclusion   At the very end all questions ever raised will be accounted for.

Theoretical Explanation of the Experimental Results

The greenhouse warming problem is essentially an energy accounting one trying to come up with a balance sheet that shows a net amount of greenhouse warming, which has been uncertain for twenty years. What is wrong with the pundits? First, they disregard the laws set up by the smoking guns and made flimflam entries away from reality. Moreover, the greenhouse heat account has gone to a first escrow account of heat of vaporization, and then to a second escrow account of heat of condensation and then to a third escrow account of heat of reflection of sunlight. Few accountants are familiar with such an involved escrow system, which is not in any textbooks and is up to the accountants to discover and unravel. Instead they are snowed under by millions of petty cash accounts that waste their time and they missed the crucial points.

That the correct balance sheet shows a zero amount of greenhouse warming is crucially hinged on the figure $16.6W/m^2$, the dissipative power of cloud. This is a number determined by God's will, free from human faking and doctoring. One cannot but wonder why God is so kind to humans to get rid of all greenhouse heat in such a generous and clean cut manner. The answer is that it is not an accident, nor a freak, but a matter of a fundamental law of physics, the Second Law of Thermodynamics, the only law that is never excepted.

This law tends to maintain equilibrium. Greenhouse warming is a deviation from equilibrium which God abhors and would do everything He could to restore equilibrium. In this case He manages to increase the cloud coverage to cool down the warming. Otherwise the earth would keep on warming up and the accumulated heat could be used to make a *perpetual motion machine* which would generate endless energy without fuel cost, which God forbid (violation of the Second Law). A condition for this argument to hold is the existence of a constant low temperature heat reservoir to act as the *condenser* (for the perpetual motion machine) which is provided for by the iced water in the polar regions (always at 0°C). This argument does not apply on Venus which has no ice ages and thus no iced polar water, which does exist on earth as a special case.

Can we count on God indefinitely? No! Good luck is rare and fleeting. Cloud is increasing at the rate of 4.1% per 50 years. After 2000 years the sky will be fully covered with cloud and there will be no more room to add new cloud to reflect away sunlight to cool down earth. Then the polar ice (a remnant of the ice age) would melt to cool down the earth to fulfill God's will. After the polar ice is gone then greenhouse warming would commence in earnest as in Venus. However, fossil fuels last only for 300 years, far from reaching the end of the world.

Thus there is an element of *luck* by which we can escape the curse of greenhouse warming for the foreseeable future. Not just fossil fuel is a triviality but that we happen to live in a time (Holocene of the ice age) the sky is half clouded (the chance of which is very small in geological history) so that the cloud can act as a *thermostat* to maintain stable temperature for 10,000 years or so. A much greater benefit than avoiding greenhouse warming is the chance for Homo Sapiens to develop agriculture and civilization which began only at the last ice retreat of the ice age 10,000 years ago. *What a good luck we got to catch just the opportune time!* Science depends on luck? Yes! But only for the most felicitous and yet most hard to reach events, such as the origination of bi-ped locomotion and oral speech which humans barely squeezed through by luck, leaving all other animals behind, and leaving all scholars studying humans behind. And all studying ice ages behind. Greenhouse warming is just a tiny bit of a much greater and fascinating climate drama centered on ice age. The details of all these will be elaborated in my forthcoming book *Two Major Energy Issues: Greenhouse Warming and Nuclear Hazard* with an Appendix: *A billion year climate history and the evolution of humanity*.

Deeper Insides and the Fallacy of Kyoto Protocol

The article could have stopped here but for an interesting connection with the current world affair. President Bush discarded Kyoto protocol, earned the ire of our European allies. He was perfectly right in saying that the greenhouse warming is still an unsettled issue, which is not challenged by any scientists and any politicians. But the evidences as we presented here are that the issue is settled to the negative and the Kyoto Protocol is a hoax. How can the European leaders be so misled by their scientific tutors from wisdom to foolishness? A super-heavy-weight argument is required to justify such an ostentatious statement to prove it beyond legal evidence, intellectual fortitude and ethical rectitude. It is because the theories do not take into account the intricacy of the Second Law related statistical physics and do not recognize the origin of ice ages (see my forthcoming book). And the experimental evidences are miss-interpreted or irrelevant mainly because of the mixed up of greenhouse *effects* with *warming*.

Beginning with the most obvious that distinguishes wisdom from foolishness—the problem of the origin of ice ages. The establishment thinking is that heat deficit produced lower temperature which then produced the ice of the ice age. The fact is just the opposite: Heat deficit produced the ice first which then produced the low temperature of the ice age. It mistakes the effect as the cause. Why? Because of the phenomenon of *phase transition* (ice to water, also water to vapor) in which heat changes the amount of ice but not the temperature—the mundane principle of ice box used in the 1930s. It is the principle underlying the working of the *thermostats*, such as the ice box referred here and the action of cloud mentioned earlier, which we will return later.

This mistake makes the theory of ice ages gone astray. For our sake this mistake is inherited in the global warming problem. The "evidences" supporting global *warming* include the receding of mountain glaciers, decrease of polar sea ice and rising sea level, etc. These involve *phase transition* with change of ice volume but not change of temperature. The latter means no global *warming*, which they prove exactly. What is actually proved is the increase of greenhouse *heat* due to greenhouse gases, which is known to everyone that needs no further proof.

This discussion also brings out the important point from the theory of ice ages that the permanent ice of the earth today in the polar regions (3×10^{22} gm) acts as an *ultimate thermostat* to maintain a stable temperature of the earth (just as in the ice box) after or together with the *cloud thermostat* mentioned earlier which is also based on phase transition with heat change but without temperature change. Thus, even though there is heat change due to whatever causes (including greenhouse heat) there is no temperature change, that is, no global *warming*. It goes without saying that anyone knows the origin of ice ages will know no global warming today. More elaboration follows.

In the ice age after the melting of the continental ice sheets (end of Pleistocene) 12000 B.P., the polar ice thermostat was too far away from the hot equator and was relegated to the "back burner," and the job of stabilizing the earth temperature falls to the cloud thermostat now in the "front burner." It took 8000 years of heating toward the end of the ice age to generate enough cloud to cover the entire earth to exhaust the cloud thermostat. And it took another 4000 years after the reversing of the ice age cycle to the cooling phase to reduce the full cloud coverage by one-half, which is where we are in today (sky half clouded). This accounts for the 12000 years of stable climate of the Holocene—the extremely long *interglacial* that has never been explained and now explained by the cloud thermostat. It leaves 4000 more years of good climate bestowed by the cloud thermostat before the commencement of the next ice age, which has been long overdue. (Scientists had advocated the peril of the imminent ice age before they discovered the imminent peril of global warming.) The point relevant to our discussion here is that the greenhouse heat of the entire fossil fuel reserves is a drop in the bucket for this long stable interglacial and the only effect it produces is to delay the onset of the next ice age by 300 years. Likewise the billion dollar splitting-hair study of global warming by the super-computer is no more than a drop in the bucket. The purpose of this discussion is to show that no-greenhouse-warming has a deep geophysical root. Nothing can shake the equilibrium established by the two thermostats in the interglacial, including, for example, whatever caused the little ice age of Europe 300 years ago.

Phase transition is ultimately related to the Second Law which can lead directly to the same conclusion as we have mentioned earlier. Not to be elaborated here is that the Second Law is the ultimate origin of the phenomena of no greenhouse warming (see my book). An important point is that the Second Law has no experimental error and leaves no room for future revision and addition of new developments. This is tremendously important to global problems which leaves endless possibilities of adding something new. In the present case two special features also help to simplify the problem: 1, the conclusion is a null solution; 2, fossil fuels have a limited lifetime of a mere 300 years.

Practically significant is the fact that phase transition is the Gordian knot of the computer solution of the global warming problem which will bug the computer for a log time. The theorists' main tool of computer solution of greenhouse warming is the *general circulation model* (GCM). It is a fluid dynamical theory based on differential equations. As such it is a *continuous* theory and cannot avoid the continuous *van der Waal's equation*. That equation has an unrealistic *ghost* solution. Conventionally it is rendered *real* by a magic *Maxwell rule* into a *discontinuous equation of state* representing *phase transition.* (How many physicists today still know the Maxwell rule?) That is something that cannot be computerized— from ghost to real is not a rational process that can be digitized.

Now the formation of cloud is a phase transition and thus the cloud is the Achilles' heel of GCM, which is well known to the practitioners. The theory cannot predict the exact cloud amount, altitude, and the water drop size of each cloud. Though meteorologists can predict rain and cloudiness in daily weather forecast, they cannot predict the optical properties of absorption and reflection of the clouds produced, of which the only thing we know is the ERBE satellite measurement of 16.6 W/m^2 mentioned in the beginning and the conclusion we draw from it is the only reliable one available today. Nineteen models of GCM predict cloud effects differing by 300% and the theory was utterly unreliable. No super-computer can unravel the Gordian Knot.

Recent experimental evidences to promote greenhouse warming are mostly irrelevant—they concern sea level, mountain glaciers, flood, etc., none with *global temperature* and its effect on pest multiplication and shift of the wheat belt, etc. These evidences do not involve the change of temperature. They are caused by greenhouse *heat* that does not involve change of temperature, i.e., not related to greenhouse warming. Flood is a result of precipitation which is a greenhouse *effect,* not *warming,* and we have smoking gun #2 to show no warming coming out of it.

As far as global temperature is concerned, that of the ocean has been determined for a century, of the Arctic sea for 50 years and for the satellite altitude for 20 years and all show no change for the study period (within the statistical error). The so-called global warming of 0.6°C of the past century is small compared with the diurnal change of 10°C and seasonal change of 100°C; it is approaching zero by all standard. The only *temperature change* evidence cited is that of the last 5 years, which show a little increase. Nothing in 5 years can change the decades to century trend. I am sure that when the 30-year-record cold of the last winter of 2000-2001 is included in the average, the 5 year increase recently would return to zero.

The recent American National Academy of Science group study of global warming, urged by the Bush Administration, agrees with the mostly European IPCC study in the trivial and irrelevant areas such as sea level, glacier ice and flood, but is critical of it regarding the *increase* of global *warming*. They reconfirm greenhouse *heat* and its resulting climate changes such as flood which no one disputes and thus add nothing new. To use this to certify *global warming* is logically mistaken and indeed scientifically wrong. The point that there is no meaningful global warming *at the present time* is well established beyond doubt—the puzzling question is the *missing greenhouse heat* which manifests in no global warming. The only remaining and critical issue is *future warming*. The NAS Panel disavows the IPCC prediction on the future warming, which is crucial.

The IPCC group contends that the future warming could go up to as high as 11°F. The argument is that future environmental cleaning would remove the polluted aerosols that have a cooling effect (the origin of the missing greenhouse heat) and after the good rogues are expelled, the currently hidden warming would then re-surface. The fatal defect of the argument is that the hidden warming is something that has never been proved before in the first place and all the world effort today is trying to prove the existence of that warming. Now they proved warming by assuming the unproved warming in the first place. This is not logical. We prove that the missing greenhouse heat is due to heat dissipation by the clouds created by greenhouse effect—by three smoking guns. There is no room left for the polluted aerosols to do anything. Further environmental cleaning has nothing to do with global warming.

With that the case is closed. President Bush has plenty of words to reply to European dignitaries, Kings, Presidents, Prime Ministers, Gorbachev and Hawking. As long as there is no *warming*, the use of fossil fuels and the increase of greenhouse gases do no harm and Kyoto Protocol is moot and unnecessary. The air pollution that fossil fuels generated is not due to warming. Carbon dioxide is not a pollutant. I repeat: not a pollutant. The bona fide pollutants from fossil fuels can be and should be solved by nuclear power which is also beneficial to health—another issue of the age of wisdom and the age of foolishness (see related articles).

The Bush Administration questioned the NAS Penal on why the Executive summary of IPCC is different from its detailed researches of three volumes, and is skewed to advocate global warming. The answer is an open secret: The three volumes of IPCC Report, 1990, are worked out by three groups in one year simultaneously—the first volume on scientific fundamentals, the second on greenhouse effects and the third on social remedies. Logically the conclusions of volume 1 should be used as the premise of Volume 2, those of 2 should be used as the premise of 3. Thus the three volumes should be worked out *in tandem*, one after another, requiring 3 years, not simultaneously, pressed into 1 year as actually happened. What came to pass was that the premises of 2 and 3 were the *assumed conclusions* before they were actually worked out in a year in the previous volume. Thus the 3 year job is pressed into 1. Any student writing a dissertation on such a basis would be summarily expelled no matter what conclusions are arrived at. But this was what actually happened and publicly announced for the purpose of saving time (from 3 years to 1)—a total disregard of logic sanctity. Needless to say the Executive summary is an updated presumption skewed to global warming, while glossing over the unsettled expert opinions. The NAS Panel criticized IPCC just in this way. Bush is right the issue has not been settled. But it is settled now in this work and in all approaches possible.

The greenhouse warming is a global issue and needs the global perspective of the global climate (especially the ice age) and the global law of all processes (especially the Second Law of Thermodynamics). Take the minor issue of sea level rising for example. The secular 2mm/year rise cannot be due to global warming, nor due to thermal expansion, nor due to icesheet melting, but can well be due to the sedimentation of the runoff of all rivers going to the ocean, which no one has calculated before, but is of the right order of magnitude. Global warming is a problem that needs Renaissance men to synthesize and explore all needed knowledge and nail down the answer, not a group of incoherent specialists with unsettled opinions. It is like building the atomic bomb which is a previously unexplored venture with a great many uncertainties. But the basic laws are known. Every issue can and must be settled without the slightest doubt and the final product be assured not to be a dud. And indeed it is not a dud—its explosion time can be controlled to a fraction of a second. The greenhouse warming is an atomic bomb that will never explode in decades and certainly will be a dud after 300 years. It is such a study that decides no global warming and Kyoto Protocol is a hoax.

This concludes another scene of the drama of the age of wisdom and the age of foolishness.

Nuclear Waste Disposal: from Trash to Treasure

Peter Fong

Physics Department, Emory Univeristy

Atlanta, Georgia 30322

The last problem bugging nuclear power is its waste disposal; the cancer risk has been behind us since the 1996 Conference of One Decade After Chernobyl sponsored by United Nations, European Union, World Health Organization and the three suffered Republics of Russia, White Russia and Ukraine, which proclaimed that there was no increase of cancer after Chernobyl.

The waste disposal has a different character. The nuclear radiation cannot be destroyed by fire, by acid or by anything. If harmful, it is forever. It cannot be diluted like a poison—the argument is that if it is not destructible and low level radiation is harmful at any dose down to zero (a point that has never been proved), then the total effect would always add up to the same without lessening.

In recent years evidence is accumulating that on the contrary low level radiation below a certain threshold is not harmful but beneficial. The most outstanding case is one in a United Nations paper on human life expectancy of Kerala, India, which is 10.7 years longer than all India, 1951-1981, because of higher background radiation (20 times) due to the large thorium mine deposits.

If the low level radiation beneficial effect is established in general then the nuclear waste problem can be solved in the most propitious way: turn trash into treasure. A deadly problem is removed and we have a windfall of extended life. Nuclear waste may be diluted to make radiation health devices as practiced in Brazil, Chekoslovakia and other places to make a big fortune with benefits to all.

Nothing can increase lifetime by ten years. Even if all cancers are eliminated, the life expectancy can be increased only 2.5 years, as recent research has concluded. And Kerala is not the only thing we know. There are much better statistics here at home—a truck load of millions of statistics available in any decent library that shows the death rates of all leading diseases of all 50 states of US in every year of the past half century. This is the ultimate data set and nothing can excel it. It even includes the detailed effects of the airborne nuclear weapons tests. I have studied it and my conclusions are: (1) The 8 mountain states with twice higher background radiation taken as a whole have mortality rates of all 8 leading diseases causing death including heart diseases, cancer, stroke,..., lower than the nation as a whole by about 25%. (2) The 8 mountain states taking separately with different altitudes, the mortality rates and the background radiation vary with their altitudes by the same exponential law, which proves conclusively that the mortality variation is caused by the radiation variation (proxy by altitude), i.e., radiation lowers mortality. The results will be published in my forth coming book *Two Major Energy Issues: Greenhouse Warming and Nuclear Hazard.* By the way the conclusion of the nuclear weapons test study verifies exactly the beneficial effect of radiation, which is a great irony.

If mountain states mortality rates of all diseases are reduced 25% at the twice higher radiation level, then it shows the live expectancy may increase with radiation. The Kerala result of 10.7 years increase at 20 times the radiation is comparable and credible. There are numerous evidence supporting the increase of lifetime by radiation (Japanese atomic bomb survivors, radium watch dial painters, high altitude village in China, high radiation city in Iran, etc.). Lifetime cannot be calculated from mortality rates. Another supporting evidence is that the menopause age of women in Denver at 2 times the background radiation is delayed by 1 year, suggesting an increase of life span of 1.3 years. By proportion Kerala women's life span at 20 times radiation should be increased by 13 years. Statistics show it is increased by 11.9 years (2.4 years more than man). The agreement is perfect.

It is extraordinary that the low level radiation is beneficial instead of harmful as usually conceived. But this is not surprising. The polio vaccine is just dead or weakened polio virus but it is beneficial to humans to stimulate immune reaction again polio. Radiation is an encroaching agent like polio virus. For life to survive on the earth it must develop an immune system to repair the damage of radiation. Otherwise natural background radiation would kill all and no live will be present today. Radiation damage is largely through oxygen free radical. The immune system against radiation is likely to be a free radical destroyer. It can be mobilized by the stimulation of excess radiation. The leading diseases of death are degenerative diseases often associated with aging and excess of free radicals. Antioxidants are generally prescribed. The immune system's free radical destroyer is just right to ameliorate these diseases and thus radiation can improve health and extend lifetime. This is just a simple idea illustrating how radiation may work.

After the Chernobyl accident thousands of European women aborted their unborn children for fear of giving birth to deformed offspring. Now it has been proved that the fear was unsubstantiated. It is surely an age of foolishness. The anti-abortionists should go after the anti-nuclear advocates that involve thousands of fetus instead of abortion doctor that involves only one. It is an example of a common psychology to mistake a high stake as a high probability. That is why people buy lottery tickets in droves—the higher the stake the longer the line of buyers. It is a stupidity and a tragedy. Worse still, the nation's energy policy was determined by this frenzy lottery psychology. During the scarce of the China syndrome of nuclear risks, the United States has aborted one hundred nuclear power plants (don't laugh at the European abortion women) and we are now reaping its bitter consequences. It is time to establish a national nuclear policy based on solid science coming out of our available adequate scientific data, which we never learn to use before and fall victim to the lottery psychology and nuclear mythology.

Without profound knowledge it is clear that 300 years later all fossil fuels would be depleted and nuclear power is the only practical energy source available in every aspect of energy policy considerations. Safety factor being eliminated, major considerations are that it is environmentally clean and economically viable. Air pollution accounts for the majority of environmental deaths, which is absent in nuclear power. Nuclear power was cheap and competitive in the beginning but the price soared after Three-Mile-Island and Chernobyl because of excessive safety measures, which is now extravagant, and the accompanying lengthened construction time from 3 years (Rochester Plant) to 14 years (presently), which is now unnecessary. The mere extra interest cost of the added 11 years doubles the *investment cost*. Together with the increased *construction cost*, the cost of electricity produced is increased 3 times, making it undesirable. With the safety question resolved the cost of electricity from new nuclear plants can be reduced 3 times. A massive building of nuclear plants is the most reasonable energy policy.

Last but not the least is the newest nuclear waste—the nuclear explosives in the destroyed ICBMs according to current and future disarmament treaties. With proper treatment they can be made into non-explosive nuclear fuel for nuclear power plant which is more efficient like that in nuclear submarine but much cheaper because it is cost free from "nuclear waste." From 10,000 nuclear warheads there is enough for 1000 nuclear plants which is enough to supply all electricity for the entire U.S. without one iota of fossil fuels. Since a nuclear plant lasts 30 years, we have 30 years of electricity free of fuel costs and free of pollution. Nothing in the world can be better.

The Shoreham Nuclear Plant in Long Island was completely built and ready to generate electricity but was abandoned. Affluent Long Island citizens would rather pay triple electric cost to redeem the loss for what was considered as safety. Poor Americans would have to go back to live in caves. This is surely an age of wisdom and an age of foolishness.

Cancer Death Rates of all States of U.S. in all Past 50 Years Related to Natural and Artificial Nuclear Radiation

Peter Fong

Physics Department, Emory University

Atlanta, Georgia 30322

Abstract

Cancer death rates of 50 States of U.S. in all the past 50 years are analyzed according to natural and artificial nuclear radiation. Figure 1 shows their dependence on altitudes of the States (perennial natural radiation)—the higher the lower deaths. The 8 mountain States average is that a doubling of natural radiation 100 mrem/yr reduces the cancer death rate 25%. Figure 2 shows their dependence on distances from Nevada, the site of airborne nuclear weapons tests in 1950s (one-shot artificial nuclear radiation of 30 mrem/yr for the test years). Nevada showed a reduction of 25% of cancer death rate after the first 4 years and continued for more than a decade. Alaska and Hawaii show no effect. The other States show reduction related to distance and peculiar features timed with the bomb explosion. The national average is 24.3% cancer rate reduction for 100 mrem/yr of bomb radiation, in perfect agreement with natural radiation.

Cancer

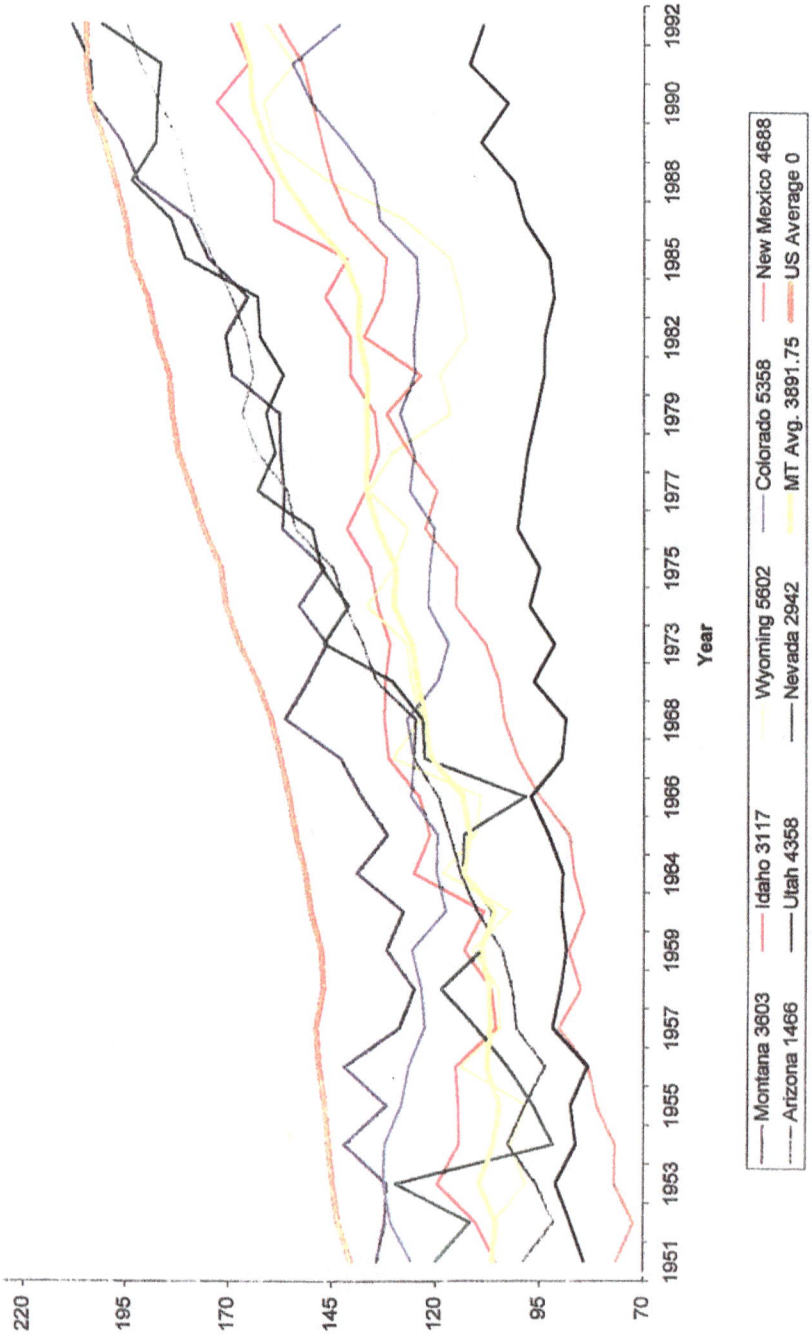

Figure 1

Figure 1 shows the cancer death rates per 100,000 population in the U.S. over the past 50 years.[1] The thick brown curve on the top represents the national average. The thick yellow curve in the middle is the 8 mountain states average, which is about 25% lower than the national average, indicating a 25% reduction of cancer deaths over the 50 years in the high altitude region, where the natural background radiation is 100 mrem/yr higher than the normal 100. The 8 thin color curves show the 8 mountain states separately, which spread out in parallel from the yellow average according to their altitudes—higher states having lower cancer. An analysis shows the cancer reduction is exactly proportional to radiation.

Correlation does not prove *causation.* Hundreds of evidences were denied, including the above one, in policy decisions. An *experiment* corresponds to the above study would require the exposure of the entire population of the U.S. to a good amount of radiation, which is impossible. However, exactly such an experiment has been done, albeit serendipitously, in the airborne nuclear weapons tests in early 1950s. It increased the background radiation level about 30% (30 mrem/yr added to normal 100).

The *experiment* was carried out without protest from the activists and without recognition from the scientists. The results were not kept as scientific data but were kept serendipitously by health workers for their routines. An excellent chance of making important discovery was buried in the mountains of government statistics. It took one more serendipity for stupid me to bump into the gold mine by accident (the whole world was even more stupid not to learn it). The result is the lower curve in Figure 2, which is nothing but the same as the top curve of Figure 1, only analyzed in the bomb test light. The great surprise coming out of it is that the nuclear weapon tests, though increased the radiation level, did not *increase* cancer deaths but instead *reduced* it by 418,000 for the entire U.S. during the test period, *verifying* the correlation study conclusion quantitatively but *proved* it by experiment.

Experts dismissed it, considering it still as a *correlation*. It took me 5 years to find out it *really is an experiment*. The test bombs were set off in Nevada at "controllable" space and time. Besides studying the entire U. S. as a whole as in the lower curve of Figure 2, there are 50 States of U.S., each can be studied as an *independent observer* with its own peculiar, unique space-time parameters and resulting in 50 independent experiments. No experiment can be more comprehensive and it is the ultimate because it includes the entire population, willingly or not, without choice and without advance knowledge of *cancer reduction*, eliminating the necessity of placebo control experiment to guarantee the sample is not skewed.

In the upper part of Figure 2 are shown the results of 10 States so studied, from the nearest Nevada to the distant Hawaii. The analysis is similar to Figure 1, which separates the U.S. total to States with different altitudes which are not "controllable" (correlation study), but now separates in States with different distances to bomb tests, which are "controllable" (Experiment). The curves differ drastically, which can nail down the cause-effect relation conclusively. Nevada stands out drastically. In 4 years after the start of the tests, cancer rate dropped 25%, better than cancer hospitals. Cancer being "incurable" by *any*, this must be due to something beyond *all*, i.e., the new radiation from the bombs. The reduction of cancers in Nevada continued for 2 decades. The far away states Hawaii and Alaska show no effect of the bomb tests. The other 8 contiguous States all show effect of the bomb tests, more or less, according to distances from the bomb test site. The hallmark teeth-bites on the curves

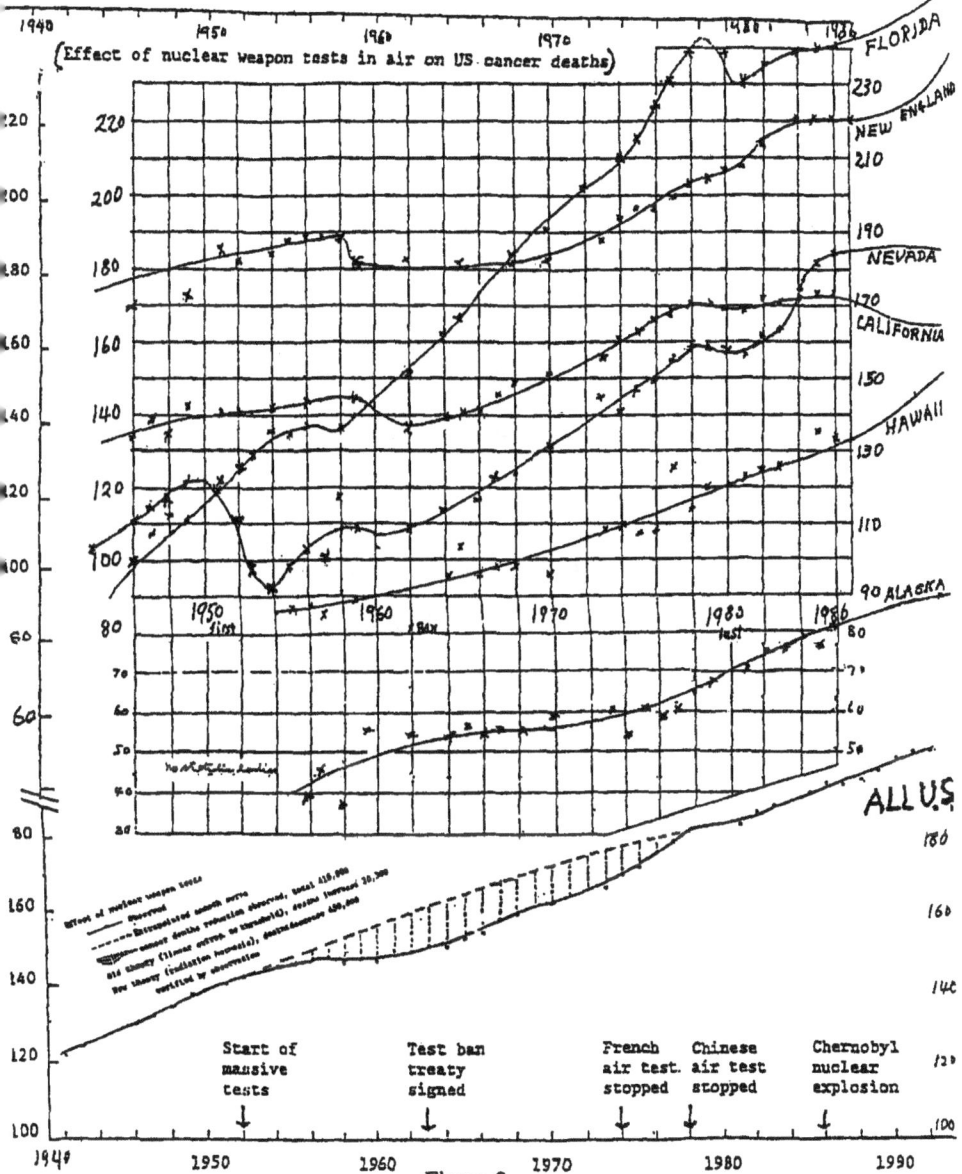

Figure 2

timed with the tests are everywhere and speak loudly for the test effect on cancer reduction. These curves prove bomb radiation can reduce cancer; the national average is 24.3% reduction for 100 mrem/yr bomb dose, which agrees extremely well with the reduction rate of natural radiation from Figure 1. Since no other study can possibly surpass this data base of 250 million persons in 50 years, the conclusion of 24.3% reduction of cancer death rate for 100 mrem/yr increase of nuclear radiation here obtained should be considered to be the ultimate.

The Nuclear Test Ban Treaty was promulgated in 1959, ten years after the start of the massive tests in Nevada, in response to the world opinion on nuclear radiation hazard. The beneficial effect was totally unexpected. Starting 1960 nuclear tests went underground and in two decades 3 times more radiation was generated but buried underground. If it were in the air it would have saved 1.5 million lives from cancer death. It is the most grotesque irony of the test bane treaty which grows out of a highly inspired humanistic ideal. Even more grotesque is that this is just the beginning of a series that characterize the 20th century—one that is the age of wisdom and also the age of foolishness.

Later we will show that nuclear radiation not only helps cancer but also all old-age degenerative diseases, including the major killers of heart disease and stroke, and thus will extend life span. Indeed 29 million people in Kerala, India have life expectancy 10.7 years longer than one billion Indians due to nuclear radiation from thorium mine. Currently nuclear plant is considered as a scourge; people stay away as if a plague. In the future it would be a benefactor; people would frock to live near a nuclear plant and ask for release of more radiation so that they can live ten years longer. The nuclear hysteria is the utmost stupidity of the 20th century, not to mention the economic losses, environmental degradation, and international crisis because of the abandon of nuclear power.

References

[1]Statistical Abstract of the United States, Government Printing Office, Washington, D.C., 50 volumes from 1948 to 1998.

Figure Captions

Figure 1. Cancer death rate per 100,000 population of the States of U.S. analyzed according to their altitudes (perennial natural radiation). The altitudes of the States in foot unit given in the figure are the population weighted averages of the municipalities of the States accurate to 4 figures.

Figure 2. Cancer dearth rate per 100,000 population of the States of U.S. analyzed according to their distances from the bomb test site Nevada (one-shot nuclear bomb test radiation).

Part I Greenhouse Warming

An essay on the controversy of greenhouse warming

Appendix: A billion year climate history and the evolution of humanity

Related research papers on the origin of ice ages

— Latent heat of melting and its importance of glaciation cycles

— Origin of ice ages: Initial condition forcing and dynamics

— Unraveling a century's mystery of the ice ages

— The fourth phase of water and the theories of greenhouse effects and ice age glacial cycles

The above is a part of the research work on environment and energy published in book (by Macmillan Publishing Co.), in national and local newspapers and magazines, in invited papers to American Physical Society and given in national, regional and local public lectures. Honors include the Key of the City of Birminghan presented by the Mayor.

An Essay on the Controversy of Greenhouse Warming

Contents

An Essay on the Controversy of Greenhouse Warming

© 2001

Peter Fong

Physics Department, Emory University

Atlanta, Georgia 30322

Prologue

The greenhouse *heat* due to increased carbon dioxide is real but the expected global *warming* has not appeared. The *warming of temperature* is one of 3 major greenhouse *effects*, which we do not dispute. Our point here is that this is the least important one, nearly zero as known today and will remain so after fossil fuels are gone. No one bother much about the other 2 effects. Since global warming has become an issue, the major scientific problem is the *missing greenhouse heat* that cuts out the warming. The difference of heat and temperature is well known to every high school student having passed physics, but is vague to laymen and the pundits, resulting in confusion. Scientists want to find the missing greenhouse heat, to make sure it will not return to make havoc and be prepared to defend it ahead of time if it should return. The laymen had the totally wrong idea that the devil is already in the house and we must put out the fire before we do anything else. That leads to the incredible mess we are in today.

We still have to search piously for the missing greenhouse heat as if the Holy Grail for the rightful scientific reason cited above and this is the detective story we are going to tell here. Science neophytes may read it as a thrilling detective melodrama without missing a hit by taking all scientific terms in *Italics* as if evidence from super-detective Lee Chang-Yu without further question. Experts may follow up from the original references given in the text in a professional way and any challenge against *Einstein's dictum* would be welcome. The conclusion is that the missing greenhouse heat is dissipated to outer space through the reflection of additional sunlight by the increased clouds created by the greenhouse *effect*. The predicted increases of clouds and precipitation of the past century are checked out correctly with observed statistics and whatever greenhouse *heat* remaining will cause a global *warming* practically equal to zero. This answers the question why there is no global warming today.

Since the reflected sunlight is short wave, it can pass through the carbon dioxide blanket which can only stop the long wave heat. This is the crucial point. The eye-catching idea of greenhouse warming is that the carbon dioxide blanket will keep the heat in but the fatal mistake is to forget that it is leaking for the short waves. The citadel wall of the greenhouse castle crumbling and there is no more battlements the warriors can fight a desperate war of defense.

It may be argued that this is just one of the scenarios that could happen but there are hundreds of possible scenarios, many have been put forward, but God has not made the final decision and anyone can insist on his pet idea to drag out an endless contest. To break the stalemate, we must prove one side is absolutely wrong. In the following we will show that within the next 2000 years the idea of *environmental greenhouse warming* is against the most fundamental second law of thermodynamics and the *Einstein Dictum* (see later), thus we have no warming to worry about in that period. I hasten to add that the 2000 years limit does not apply to the historical, *natural greenhouse warming* that made earth warm enough to harbor life, which is well known (see the Appendix). The crucial point is that the present earth is half-clouded and partly-iced as an historical heritage of the *interglacial* earth and thus its *climate* is largely controlled by the *three-phase equilibrium of water* (see the Appendix) rather than by *fluid dynamics* as in *weather*. The critical fact is the existence of a permanent ice in the polar regions today.

The origin of *permanent ice* on earth today is the *ice age*, an important problem by itself and in planetary evolution, and in the emergence of humans on earth, and the problem of population of intelligent beings in the universe, all important, related and interesting problems that will be briefly touched upon in the Appendix. The main text will be limited to the thesis of the absence of anthropogenic global warming now in the interglacial earth, the past being relegated to the Appendix. The Kyoto Protocol is *moot* because any certified warming will only half-way reach the hazardous level before the fossil fuels will be depleted in 300 years, far short of 2000, then carbon dioxide will stop increasing and there will be no more such warming forever after.

IPCC Update

The *Intergovernmental Panel on Climate Change* (IPCC) has recently released information (New York Times, Oct. 26, 2000, P. 18) of the latest research "close to consensus" that the earlier 1995 vague prediction of possible global warming—*first glimmer* of greenhouse warming seen (as reported by *Science*)—is now clearly out of date, suggesting certainty without saying it. Dissident MIT Professor Richard S. Lindzen contended that the new study is just to keep the issue alive and confirmed that there is no evidence that global warming would have harmful effects.

Scientists were happy to find one possible answer to the missing greenhouse heat—the *polluted aerosols* reflected away sunlight and cooled down the earth. After many years of no further warming, the report declared that warming will sure to come after environmental cleaning later has removed the polluted aerosols. We have been anxiously awaiting a brilliant breakthrough, not a lucky strike or a cunning sophistry. Actually the *good deed* is done by the *increased cloud*, shown by *cloud, precipitation* data (later), and will not be undone by cleaner air. The predicted warming of 1.4-5.8°C is a mirage, a clumsy childish lie to fool you all. As a childish lie it is innocent but deplorable, with no malice but with great calamity.

It has been said that the extreme warmth of the past 5 years is a clear proof of the *coming* of global warming. But then the record breaking cold of this winter (2000-2001) worldwide is a proof against global warming for violating energy conservation. See La Niña later for the recent cold.

The IPCC results will be proselytized in a series of UN sponsored world meetings from Shanghai (January 2001) to Nairobi (April 2001) in a program parallel to the IPCC 1990 Report with updatings as revealed in New York Times in October, 2000. The results will spearhead the Seventh UN Climate Protocol Meeting in Bonn (May 2001) to rescue the failed Sixth Meeting in Hague to reduce the carbon dioxide release of the wealthy nations (see at the end of the main text). The charge brought up here that there will be no environmental greenhouse warming in the next 2000 years must be absolved lest the much ado becomes the joke of the century.

The Counterpoint: The Second Law of Thermodynamics

There are more dissidents than popularly known. During the 1996 Nobel Centennial Debate in Stockholm, Nobel Laureates openly debated on the viability of the idea. In April 1998 Frederick Seitz, former President of the American Physical Society, Rockefeler University and National Academy of Sciences, circulated a petition against Senate approval of the Kyoto Protocol designed to stop global warming. Books, Conferences, Video tapes in the opposing camp abound. But the media and the establishment pay no attention to the dissident opinions and pursued their self interest, just like in the Lee Wen-Ho case, until the judge discovered it and publicly reprimanded them, and apologized to Lee. And the American public awakened that they were being fooled. Indeed, all the time. See New York Times, Feb. 4 and 5, 2001 for an exhaustive, shocking reconstruction of the flopped case against Lee.

It is time to clear up the air. I have stated the thesis that the current idea of greenhouse warming violates the second law of thermodynamics, making it possible to make a perpetual motion machine, which would generate endless electric energy without any fuel costs. If the greenhouse heat of the *atmosphere* is used as the *boiler* heat and the polar ocean as the *condenser* (its surface temperature is always kept at 0°C because of the polar ice) to operate a *heat engine* (see any freshman physics), it would indeed be a perpetual motion machine that would generate endless energy.

The temperature in the farmer's greenhouse is indeed jacked up, but is sealed in an air-tight glass enclosure, just as in the roof-top solar heater, a low-tech device that does not violate the Second Law. On the other hand the greenhouse heat of the earth atmosphere, according to the advocates, is supposed to be sealed by a carbon dioxide blanket in the atmosphere which prevents it from escaping the earth. It follows that the temperature of the atmosphere can be jacked up and it can be used as the boiler heat, together with the polar ocean as the condenser to make a perpetual motion machine to generate endless electric energy, which violates the Second Law.

If the blanket is indeed made of an insulating wall acting as an enclosure like that of the farmer's greenhouse, all that has been said of the greenhouse effect according to the advocates would be true and perpetual motion machine would be realized indeed. *Hurrah!* However, the blanket is made of a layer of gas of carbon dioxide, which every kid knows is full of holes. It is through these holes, communication can be made with the world outside by a mechanism of *three phase equilibrium* of water to be detailed later that the seemingly impenetrable blanket becomes an empty net and nothing can be corralled inside—no global warming at all. It is not an issue to be determined by consensus, to be decided by expanded research, or can be changed in 5 years. It is determined by the most fundamental law of the universe and the *interglacial* status of the earth today. Any compromise of it is as wise as promoting anti-gravity material to make a flying carpet.

I have given lectures, published papers on it without receiving rebuttals and supports; and decided for a public open confrontation. The occasion was the April 1994 Washington meeting of the American Physical Society, in which a special Invited Paper Session was devoted to the subject. After the first paper delivered by S. H. Schneider I raised the question how to reconcile the greenhouse effect with the second law of thermodynamics which would make it possible to make a perpetual motion machine of the second kind. He relinquished rights for rebuttal to other experts and M. Hoffert and Michael MacCracken answered the call. The gist of the rebuttal was that the ocean surface temperature is not stable due to surface winds and so on, so that it cannot be used as the *condenser* of the perpetual motion machine. The speaker did add, to show scientific neutrality, that a few meters below the ocean surface, the temperature would be stable at 0°C not affected by everything including greenhouse warming. This was a godsend to me—like the football player running in the opposite direction and scored a touch down for the opponent in the Rose Bowl Game. Just put the condenser a few meters below the ocean surface and you would have a perfect perpetual motion machine that could provide free electric energy from the greenhouse heat and avoid global warming at the same time. This would be truly a flying carpet.

An Alternative Approach: The Le Chatlier Principle

While the argument is clear cut for anyone having had a freshman physics, another argument is more palatable for those who have had a freshman chemistry or just science for dummies (still good, like the smart camera for dummies). That is by way of the Le Chatlier principle which is a special case of the Second Law. That principle says when an equilibrium system is disturbed by an intruding influence, the reaction of the system has an effect to remove that influence. As is well known, the wording is vague and disputable. But whatever truth coming out of it is based on the Second Law, and can be quoted without reservation. To illustrate, when a flask of reactants is heated up, then *endothermic* reaction will take place to absorb the heat and remove it, and temperature will not be increased.

This is a way suitable for us to discuss the greenhouse problem. The earth was in equilibrium. The intruding influence is the human generated greenhouse heat. And the effect is *something* to take place to remove the heat and the temperature will not be increased. Thus no greenhouse warming. It appears as if equilibrium is God's will, and any intruder, such as the ugly human, trying to violate it, God will send something—a holy thief to steal the thunder of the Devil and return peace to the earth.

Where Goes the Missing Greenhouse Heat?

God's will is not to be taken for granted until it is demonstrated by God's hand—to ferret out the holy thief that has stolen the greenhouse heat. I gave an open book test asking students to "Argue that the second law of thermodynamics forbids global warming."

The first student argued that the melting of the polar ice would remove the greenhouse heat and cool down the earth. This insight is not from God's will but from common sense—the 1930s kitchen ice box and today's ice cubes in the orange juice (actually the Le Chatlier principle). I gave him a grade of 60/100 for arguing in the right direction but the answer is wrong. Melting of polar ice would raise the sea level but in reality the sea level is constant. To be exact the sea level rises a mere 2mm/year but would be 35 times more, that is, 7 cm/year, if polar ice would melt by greenhouse heat. The thief has not been caught (it actually has stolen others and prevented the polar ice from melting).

The second student argued that the greenhouse heat was consumed in evaporation of the ocean water, which then cooled down the earth. Again it is not God's will but common sense. It is the 1930s cooling machine that blows air over a dish of water to evaporate and to cool down the room (Le Chatlier again). I gave him a grade 70/100 for arguing in the right direction and better than the polar ice argument—ocean covers most of the earth surface and all the warm areas whereas the polar ice is just the opposite. Yet the answer is again wrong—the sea level would *decrease* and we have already said it is *constant*.

The third student argued that the ocean water evaporated returns to the earth by precipitation, again common sense, so that the ocean level will not be lowered (Le Chatlier again). Good! And he further studied the problem in quantitative detail by digging up the latest world precipitation data of the past 100 years compiled by Oak Ridge National Laboratory, which was ignored by climate researchers (it appeared after the IPCC Report of 1990), that shows a 7.8% increase of precipitation in the past century. He further showed that the amount of *latent heat* for this amount of increased precipitation is about equal (within 7%) to the total amount of greenhouse heat dumped into the earth but removed by the evaporation of ocean water. This is pretty close to a scientific answer—quantitative verification. The job to obtain a 7.8% change in precipitation is much better than the wishy-washy 0.5% job on the greenhouse temperature change with an uncertainty (about 100%) comparable to the prediction itself ever offered by IPCC. I gave him a grade of 80/100 for the excellent job. But it still does not solve the problem. The greenhouse heat cleansed during the evaporation is re-released back to earth during the precipitation and is still stuck in the earth, now at the cloud level where vapor condenses to form water droplets. The thief has not been caught, but the student did find an important evidence that may help break the case—the thief's loot has been found to be spirited away from the sea level to the cloud level.

The road is now paved for the fourth student to break the case. Cloud! Cloud! Cloud! That is where the money is laundered and no trace of the crime is left on the earth. Cloud is known to cool down the earth by reflecting away the sunlight. You feel cool when a cloud move over your head. (Cloud could absorb sunlight and warm the earth but on a global average the net effect is cooling.) Again common sense without relying on God's will. The greenhouse heat, after the thief has scurried with it to the cloud level, has no choice but be got rid of the cloud by the cooling effect of the *increased cloud* through the reflecting away of additional sun light. Thus God's will is finally fulfilled automatically. I gave him 90/100.

The fifth student completed the proof by digging up the statistics of increasing cloud coverage of the earth in the past 50 years; its original document is included in the reference list of the IPCC Report of 1990 but not mentioned nor discussed in the text. Its conclusion is that the cloud coverage of the earth has increased 4.1% in the past 50 years. The *power* of cloud energy dissipation (cooling) is 16.6W/m^2 according to the ERBE satellite measurements. From these he calculated the *power* of cloud-increase energy dissipation for doubling of CO_2 to be 3.98W/m^2 which compares well with the theoretical value 4W/m^2 of greenhouse warming for the same period, which climate scientists concluded, with an error of 0.5%. *Thus the heat generated by greenhouse effect is all removed* (within 0.5%) by the *increased cloud* created by *this greenhouse effect*. And the *greenhouse warming* is exactly zero within 0.5%. I gave him 99/100.

Why one point short of full mark? The work provides smoking-gun proof sufficient for the jury to deliver death sentence to the perpetrator. There is no reasonable doubt to establish the guilty conviction, which can be agreed upon by all reasonable humans. But there is always unreasonable jury who insists on unreasonable standard to dissent. The scenario worked out is reasonable and quantitatively verifiable all right but has not been proven to be the only scenario available and the only one God must choose.

The answer comes from *irreversible thermodynamics* which deems that the path toward equilibrium is the one that makes for *maximum entropy production*. The legal book can now be closed with all conscience. Without going into the details of theoretical physics, we present a pedestrian argument to bring out the essence of the basic principle involved.

The Ultimate Question: Can the Second Law be Violated?

Still, God is inscrutable, as St. Augustine argued in the case of the rape of the nuns by the barbarians. There is always doubters, such as Descartes, who doubted him self's existence. Why must the Second Law and its corollary the Le Chatlier Principle be holy inviolate?

Towering scientists, such as Einstein and Eddington, believed the supremacy of the Second Law—the Word was God. However, as always, there are dissidents and heretics. There is a school led by Professor Ilya Prigogine who contends that the Second Law is not supreme but is valid only for a selected group of Hamiltonians in Newtonian mechanics—the Word was with God, but was not God. (See the Proceedings of the Pittsburgh Symposium on Thermodynamics, Mono Books, 1970). The Leiden physicists including the celebrated Ehrenfest and Crammers castigated the Second Law and considered it actually is consisted of three independent laws, the third of which, on the *increase* of *entropy*, has never been proven in the *active* sense. I was one of the rebels, because I do not believe in absolute authority, which corrupts absolutely. And I smelled something fishy, like the sale of Indulgences, which suggested that there might be indeed a Hamiltonian that violates the Second Law, and I concocted a perpetual motion machine to demonstrate it. The description of the machine can be found in my articles on Thermodynamics and Entropy in McGraw-Hill Encyclopedia of Physics. See Figure 1.

Briefly the machine consisted of two boxes containing ionized gases connected by a curved tube and subjected to a magnetic field which selects a group of high velocity particles to move from one box to the other, which cannot move backward because the Lorentz force is reversed when the velocity is reversed (time-irreversible Hamiltonian?). Thus the receiving side of the two boxes will have a higher temperature than the

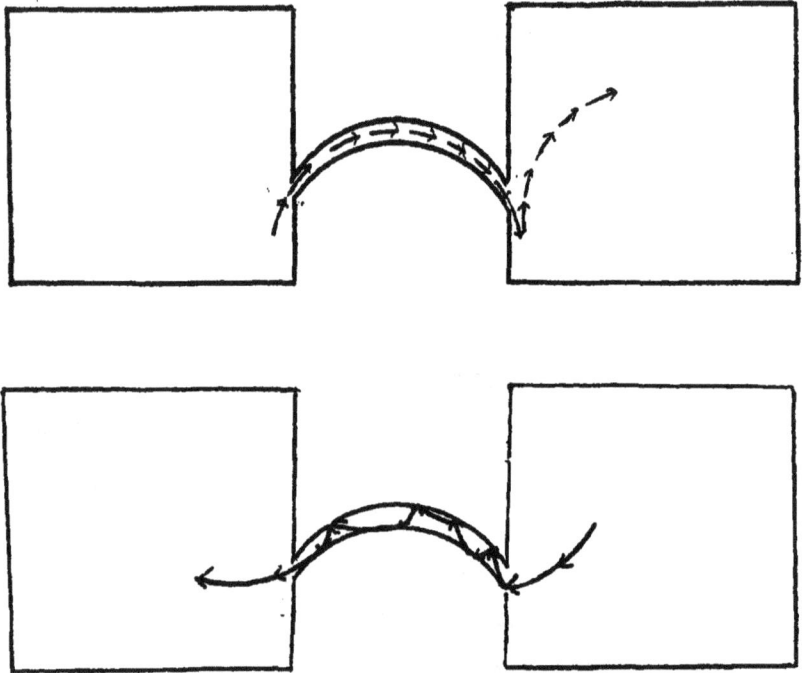

Figure 1. Fong's magnetic perpetual motion machine. The curved ion trajectory cannot be reversed, thus destroying detailed balance, seemingly working as a Maxwell Demon (upper figure). However, ions can zigzag back (lower figure) and restore gross balance. Efficiency of the machine is exactly zero (Belinfante). Sunlight comes into atmosphere and changes into greenhouse heat; it can then be changed to heat of vaporization, and then heat of condensation in cloud and then turned into reflected sunlight from cloud and go back to space, all leaking through the carbon dioxide blanket. The net greenhouse warming is exactly zero.

losing side and a perpetual motion machine can be run between them to generate electricity without any fuel costs. Bravo!

It was a sensation! It was like a Gordian Knot no one knows how to unravel. It had mesmerized all physicists including all Nobel laureates I knew except Eugene Wigner. Among them T. D. Lee had discussed it with C. N. Yang and told me that there was some faulty point in my argument for the Gordian Knot to be invincible but did not unravel it. Lee did emphasize that the *kinetic theory* approach, which I employed, was very, very complicated, hazardous and extremely prone to mistakes. The point was well known to me—it is almost impossible to derive the Clausius-Clapeyron equation by the kinetic theory even for 100 years whereas it can be derived by thermodynamics in a few minutes, but I have never gauged the intricacy and intrigue of the enterprise on the way.

Wigner wrote me a long letter, convinced of the fallacy of the contraption but did not unravel the Gordian knot. He later said he learned thermodynamics from Einstein, and Einstein said only a time-dependent Hamiltonian can destroy the Maxwell-Boltzmann distribution. This is the *Einstein Dictum*. (There is no such Hamiltonian in the real world, which would violate the energy conservation law.) It implied that the Second Law is supreme overall.

The Battle of Armageddon

Came Professor Frederick Belinfante, having moved from Leiden to Purdue, who saw a chance to explore the Leiden tradition and regarded the Fong Machine as demonstrating a Hamiltonian that might not be corralled by the Second Law.

He pursued a detailed *kinetic theory* calculation trying to confirm my qualitative argument that appears obviously correct to any descent physicist, and tried to calculate the *efficiency* of the Fong Machine for practical applications (!). It was indeed an ambitious and formidable mathematical project. To begin with he used a crude approximation and obtained a positive energy gain of the magic machine. Encouraged by it, he improved the approximation and again obtained a positive gain. Further encouraged he pursued higher accuracy approximations and obtained additional encouragement. Finally he accomplished the *exact solution* without any approximations. The result was a great surprise, the energy generated by the Fong Machine turned out to be *exactly zero*. The Second Law has triumphed at last. Wigner's prophesy and Lee's premonition were entirely fulfilled. *Ah! la maledizione!*

The greatest shock coming out of the exact calculation is that the precision marching rank and file going through the curved tube under the magnetic field *is exactly* balanced off by the back current formed by the army of drunkard stragglers bumping aimlessly on the tube walls and managed to get through backward by chance (see Figure 1 bottom), which is completely out of the common sense (*prima facie* evidence failed the meticulus *rebuttal*). It is like a recount of a million ballots coming out of sober and drunk voters resulting in half million yes votes all from sobers and half million no votes from drunkards without a single vote difference. That was why the Fong Machine was such a mesmerizing Gordian knot. But that is unmistakably indeed the *God's will*—the final result of a complex process determined by the universal law of probability controlling the process of statistical equilibrium. After all, the Word *is* God.

The crust of the matter is that the magnetic field destroys the time reversibility of the particle trajectory and thus destroys the *detailed balance* but still preserves the *gross balance* because, the Hamiltonian is not time-dependent, according to Einstein. Many earlier discussions of *statistical mechanics* were often based on *microscopic reversibility* and detailed balance. The Fong Machine was designed to challenge statistical mechanics by removing these ideas that may not have universal occurrence. The facetious success of the challenge eventually turned out to be unfounded.

Once the veil of mystery is lifted, the Gordian knot that has fooled one generation of physicists can be unraveled readily by Einstein's dictum. The two boxes are connected (not isolated) and thus there is only one final equilibrium state given by one Maxwll-Boltamann distribution, which gives one and the same temperature for the two boxes. The curved tube and the magnetic field (not time-dependent Hamiltonian) cannot destroy the Maxwell-Boltzmann distribution and cannot change the temperatures of the two boxes. Thus the Second Law is upheld. Besides establishing a significant point in basic physics, it implies the absence of global warming.

Belinfante refused to publish his extensive calculation work of many months, lest he expose himself as a stupid idiot challenging the holy Second Law. I agreed and I never mentioned the Fong Machine in my book *Foundations of Thermodynamics* published by Oxford University Press, which has been considered by many as a classic, and won high praise from many including Belinfante for correcting the shortcomings of conventional thermodynamics, which the Leiden School has long criticized.

The Sophists' Distortion

In academics nerds are permitted and encouraged to pursue unpopular ideas, some of them actually ends up as great breakthrough, such as Barbara McClintoch's work on the *jumping genes* that led to the flourishing field of genetic engineering. (She said she would have been fired if she worked in any university but she finally won the Nobel Prize.)

The scientists of greenhouse warming worked within the bounds of scientific honesty and decency as seen in IPCC's wishy-washy wordings.

But the idea was picked up by the media and activists. What the scientists considered as the *first glimmer* of greenhouse warming was blown up as an atomic bomb thunderclap, and the end of the world gospel was spreading like a wild prairie fire. They used headlines to promote an agenda that is outside the bounds of scientific honesty and decency, which could result in great harm. After 20 years of advocacy without seeing the catastrophe appearing, the promotion has become a stale, rancid joke. The Kyoto Protocol was promulgated, setting up standard for nations to follow, which is the greatest screw-up of the 20th century as will be shown shortly.

Proof by the Einstein Dictum

The greenhouse warming mechanism is actually like the set-up of a perpetual motion machine, the impossibility of which is cinched by the Le Chatlier principle (a click of the dummy's camera and you have a perfect picture) and my students' work in the open book test. If these proofs of the absence of global warming are not high brow enough, then let me prove it by the highest brow—the Einstein dictum to remove any remaining doubt.

Since the introduction of the *Maxwell Demon* it is clear that any effective *one-way mechanism* can lead to the construction of a perpetual motion machine and thus destroy the Second Law. The carbon dioxide barrier is in effect a one-way mechanism—sun energy can come in but heat energy cannot go out—thus it is an *inanimate Maxwell Demon*, which would be the most startling discovery in science ever, not to mention that a perpetual motion machine can now be constructed. In history a dozen one-way mechanisms have been proposed to do so, but can be proven to be unworkable by microscopic reversibility of mechanical processes.

To probe this point further we note the *ratchet machine* as a one-way mechanism—the car can be jacked up but will not go down—that might lead to a perpetual motion machine. Professor Richard Feynman has proven it impossible in a whole chapter in his general physics textbook. I have proven it impossible in one paragraph by invoking the Maxwell-Boltzmann distribution. (Also by Einstein's dictum because the ratchet is not based on a time-dependent Hamiltonian.)

The greenhouse warming system is but a more complicated, sophisticated ratchet machine trying to jack up the temperature, which I want to prove impossible by showing it does not act as a Maxwell Demon that can violate the Second Law by the most profound viewpoint developed above. Now all current calculations of greenhouse warming increase of temperature are approximations like Belinfante's approximations in the Fong Machine calculations and all such calculations showed some success of increase of temperature to violate the Second Law. But Belinfante's final exact calculation showed the increase of temperature is exactly zero without violation. The Maxwell Demon vanishes as soon as exact knowledge appears. If climate scientists worked out the exact solution, the resulting global warming would be exactly zero just as Belinfante has done.

The exact solution would include an exact solution of the simultaneous differential equations of the entire earth climate system, including the atmosphere, hydrosphere, terrestrial sphere, cryosphere and aerososphere, in fact, the exact, complete solution of the entire meteorology, oceanography, geology, glaciology and the cloud physics combined and including all their interactions, not even 1% of which is available today. Any super-computer in 100 years cannot solve it just as it cannot derive the Clausious-Clapeyron equation.

Indeed the current *general circulation models* calculations in climate theory on greenhouse warming are only reasonably successful and all 19 models are in agreement only when the atmosphere *alone* is concerned. When the *aerososphere's cloud effects* are included, all 19 models differ in results by 300%, indicating the inadequacy of the presently-known knowledge in this field.   The cloud effect is like the higher order approximations neglected in the early Belinfante calculations, which manifested as a phantom of the Maxwell Demon. But his exact solution shows the efficiency of the machine was exactly zero, 0 ± 0. The same would happen when the *exact solution* of the greenhouse machine is worked out. And my students have worked it out closely to be $0\pm0.5\%$. *Hallelujah!*

The difficulty of the cloud effect is that it involves a process of *phase transition*, which involves a *discontinuity* from a curve to a straight line, which is difficult to handle in theoretical derivation, and, needless to say, in computer programming for the solution. In theoretical physics it has engaged such celebrated physicists as Lee, Yang and Uhlenbeck. In atmospheric physics for climate studies it has been the Achilles' heel in *modeling*. It is not surprising that it played a naughty role in the greenhouse problem.

On the other hand phase transition is a point that the Le Chatlier principle has a noted specialty to handle besides its handling of chemical equilibrium.   The Maxwell Demon has no place to hide. From that principle the absence of global warming comes out in one sentence—all the complications are by-passed. All such studies are based on the Second Law on the most broad base of the Einstein dictum. This completes the proof.

Afterthoughts

The only loophole to make the cloud cooling mechanism defunct is for the increasing clouds to cover the entire sky which would happen 2000 years later at the current rate of cloud spreading. Even then, without the cloud cooling, global warming would not commence, because the ice would begin to melt first to cool down the earth until the ice would be gone completely. Thus the scientific background of no global warming is the co-existence of ice, water and vapor on the earth in a *three-phase equilibrium of water* mentioned earlier.

However, the fossil fuels lasted only for 300 years, far less than 2000, far below the critical point. The mistake of believing in global warming is not the philosophical reasoning but an error in historical judgment of the timing. In fact, the historical *natural greenhouse warming* did occur under the fully clouded sky and ice-free condition of the earth, which will be returned to in the Appendix, without the co-existence of the three phases of water.

In the meantime, within the lifetime of the fossil fuels, no matter how the pursuers of the Holy Grail of greenhouse warming belabored and no matter how the billion-dollar super-computer will predict the increase of temperature, there will be always *something* outside the computer that will siphon off the heat and bring temperature back to normal just as has happened in the higher order terms neglected in Belinfante's earlier approximations. The seemingly one-way mechanism notwithstanding and the *detailed balance* being lost, the *gross balance* is still accomplished ultimately by the Second Law and Einstein's dictum for the earth in the interglacial state.

And that *something* is the real *greenhouse effects*, which the advocates have ignored, such as *the increase of* evaporation, precipitation and *the increase of* cloud coverage, (and hypothetically the melting of polar ice which is the ultimate loot of the holy thief, but is currently interloped by the cloud and the precipitation). These form the underground railroad to let the greenhouse heat sneak through the carbon dioxide blanket, just as the curved tube in the Fong Machine provides *at the same time* a back leak to redress the imbalance. The miraculous feat is finally accomplished by the laundering of greenhouse heat from heat waves to short waves so that it can penetrate the carbon dioxide blanket. (It does not violate the entropy principle just as the drunkard stragglers balances out the rank and file marchers.) It works by changing the greenhouse heat to heat of vaporization, then re-releases it in cloud formation, and then at the cloud level, changes hands into the reflected sunlight from the increased-cloud coverage, which, being *short waves*, can leak through the carbon dioxide blanket—this *trump card turns out to be a dud* and cannot cause greenhouse warming. Thus answers the question why the *greenhouse heat is missing*.

The greenhouse heat, like a sage, comes from the dust and returns to the dust, but leaves a spiritual legacy on the earth—increased precipitation and cloud coverage without a material idol *heat* for believers to awe and worship. Unlike Belinfante, the general circulation model cannot treat the condensation process, and thus can only predict more *glimmers* of greenhouse warming without proving it as Belinfante can do.

Aside Warming what are the Impacts of Greenhouse *Effects?*

As to the legacies, *precipitation*, one of the real greenhouse *effects*, is much more hazardous than the much ballyhooed *global warming effect*. The *flood* so generated caused thousands of deaths a year, whereas the warming does not cause one single death in the worst scenario. The advocates are delinquent of their duty in promoting their cause. The increasing precipitation as a hazard of greenhouse effect should be opened for discussion. Omitting the details and jumping to the conclusion, it is more beneficial to humans in increasing the supply of fresh water for sustaining the industrial society and reducing the wild forest fires. Flood is a preventable evil (actually it is God's will in creating the fluvial plain which nourishes humans).

Another legacy, an environmental issue that is overlooked, is the impact of the *increased cloudiness* which include the following: 1. the reduced solar energy input and the reduction of agricultural yields, 2, the reduction of solar *UVB radiation* and the reduction of skin cancers that may make the *ozone depletion* issue moot,..., and finally the abrogation of the right of sun-bathers' exposure to the sun (eventually one has to take a space shuttle trip to outer space to take a sun bath). The assessment of economic and social impacts of greenhouse *warming* by IPCC is completely off the mark, belaboring on triviality and non-occurrences, and neglecting the essence and urgencies.

The recent winter cold is not a proof of the absence of greenhouse warming but one of its sophistry and the naiveté of its believers. Any changes in 5 years is not of greenhouse origin, which must appear gradually over 100 years. It must be due to other natural factors, such as El Niño (peaked in 1998, warming), La Niña (1998-2000, cooling), and solar activity changes. Since the weight of the air column above 1 cm^2 of ocean surface is equal to 10 meters of water column, the warming and cooling of ocean surface in El Niño and La Niña are sufficient to explain the corresponding changes of the climate.

As far as *warming* is concerned there is no need of any further calculation and any further experiment. The billion dollar super-computer program is a sheer waste of money and time because it is just a better approximation, not the exact solution, which is not helpful and indeed misleading. Why mount a Quixotic charge to the Einstein windmill? Have you seen the bodies of the vanquished warriors strewn in the field?

The Ozone Hole Problem

The ozone hole problem, an environmentally important issue, though outside the field of the present discussion, is related closely to scientific discoveries made here, and it would be greatly remiss not to take advantage of this opportunity to explore it. The problem is complicated and involves 1, the generation of ozone in ionosphere, 2, the appearance of chemical CFCs (freons, etc.) and *low temperature chlorine chemistry* in stratosphere that reduces ozone and creates the holes, thus increasing UVB (ultraviolate biological) radiation, and 3, the passing of UVB radiation through troposphre and city atmosphere to reach humans to cause skin cancer.

The best place to start is the second stage where we know best: The scientific accomplishment has won a Nobel Prize deservedly. Experimental proof is established by the earth satellite observation (SAMOS) that the world ozone level has been decreasing at the rate of 2% per decade.

There are other problems left in the first and third stages. We have learned in our study that the cloud level in the troposphere has been increasing at the rate of about 1% per decade which will reduce the UVB radiation by 1% per decade, reducing the CFCs' effect by one-half. Thus the experimental foundation of the ozone threat theory, 2% reduction in a decade should be changed to 1% per decade, which is hardly threatening and may well be accounted for by statistical error, not by CFCs.

In the cities the air pollution due to the very same ozone but from a different source, now the automobile exhaust, has been *reducing* the UVB level steadily in the past century, with effects far outweighing that of CFCs. No reduction of UVB in 8 big cities 1974-1985 [Science **238**, 762 (1988)]. No statistics of the skin cancer death rates of the past years have ever been made known—an important missing link in any study of this type. If any, it is likely to be limited to the unpopulated areas, which is a small percentage of the total population. It is far from a serious practical concern now. The main worry is a fundamentalist's concern of preserving a perfect environment, lest the ozone hole expands to cover the entire earth.

The crust of the matter is that the Nobel Prize tells us how the ozone *depletes* but no one knows how it *recovers*. It does recover, periodically, and in synchronization with the weather cycle—ozone disappears in winter with the appearance of the hole and reappears in spring with hole recovery.

Why? Because in warm spring the *cold chlorine chemistry* that won Nobel Prize does not work and so also the mechanism of ozone depletion. Thus ozone recovers by the mechanism of *regeneration* which was a problem overlooked and the solution of which was claimed to be unknown at the American Geophysical Union Meeting's special session celebrating the awarding of Nobel Prize for the ozone research.

The regeneration of ozone is a problem left uncompleted in the first stage on the creation of ozone in ionosphere. Ozone is not like diamond that is created to last forever. Ozone is unstable, known to every high school student. In ionosphere ozone is created by solar energy and then disappears by its own instability in a dynamic equilibrium resulting in a stable density in a balance. When the anthropogenic CFCs intervene, the equilibrium density of ozone shifts to low in winter and recovers to normal in spring, thus the coming and going of the ozone holes. The destroyed ozone by CFCs is not the lost diamonds that are lost forever as many had feared that the CFCs would eat up all ozone of the earth. It can and must regenerate and the cycle is geographically confined in the cold part of the atmosphere (the two poles) and will never intrude the populated areas. Thus it is not a threat to humans and the costly program to save ozone is absolutely not necessary.

It is a matter determined by chemical equilibrium, which is ultimately based on the second law of thermodynamics. Good Heavens! It is a *deja vu* of the global warming issue—a run-away catastrophe is averted by the same holy law.

If the ozone cycle is weather determined, then the ozone hole phenomenon is just a weather storm in upper atmosphere, not different from a hurricane in the ocean, which we consider as an act of God and pay no further attention to it. The fundamentalist's worry of ozone depletion is like worrying a tropic storm to engulf the entire earth. Both are geographically limited by nature as local and ephemeral interruptions, not everlasting catastrophe. The poor used-car owner spending $1000 to change the automobile air-conditioning system to reduce CFCs' use to save the precious ozone is pouring the money down the drain for nothing.

Of special interest is the existence of an ozone hole in Tibet, considered as *the third pole of the earth* because of the massive ice on the Himalayan Mountains. The current explanation is that the hot air vertical current from the Tibet plateau in the summer blows away the ozone and creates the ozone hole in the summer, not winter, in Tibet as it actually happens. Tibet has a population of 1 million and modern medical services for 50 years. It is incumbent to *Academia Sinica* in Beijing and the government of China to tell the world the *increase* of skin cancer death rate in Tibet over the 50 years compared with other regions of China to settle

the major question of the ozone issue for the benefit of the entire world—
to bring to light the mysteriously missing coffin in the pompous funeral.

The Kyoto Protocol

The Kyoto Protocol on reduction of fossil fuel consumption needs
further discussion. It is promoted, not by the wishy-washy scientists, but
by the self-righteous, determined activists pursuing a pre-ordained mission,
with a rigidly set agenda of social surgery intended to prevent a dormant
disease from exploding many years later.

They do not care any of the new greenhouse gases and any of the
heat dissipating effects. All they know is carbon dioxide and fossil fuels.
Limited to that framework, granted that the observed warming of the past
hundred years of 0.5°C is due to fossil fuels. The greenhouse warming in
the next three hundred years, by what they know, would be 1.5°C, which is
only half way to the threshold of 3°C increase, below which the warming
will show no ill effect, according to Schneider (Nat. Geog. Soc., Research
& Exploration, Spring, 1973, P. 173). Now three hundred years later the
fossil fuels will all be depleted and, by all they know, there will be no more
carbon dioxide increase thereafter. From this logic, the world would then
be free from the curse of greenhouse warming forever. So why surgically
remove the feet and hands to prevent the outbreak of a gangrene that never
will be in the first place? It is the most outrageous surgical screw-up ever.
It exacerbates the energy crisis, increases the risk of oil wars, slows down
the advance of the developing nations, disrupts the economy of the
developed nations, and eventually will risk a world economic depression.

The recent (Nov. 14-27, 2000) UN Climate Protocol Sixth Meeting
in Hague squabbled on the issue of *carbon dioxide sinks* (if they can reduce
the quota of carbon dioxide reduction of America, Japan and Canada, etc.).

The sink shows the complexity of the carbon dioxide problem. As far as the Kyoto Protocol is concerned, carbon dioxide seems to be the principal topic and global warming is forgotten, like a pompous funeral without a coffin. One delegate not toeing to the fundamentalist's line was smacked with a pie on the face. The whole affair of the Kyoto Protocol is a ludicrous farce! A cult antic! When will the pie turn to a bullet?

Epilogue

As far as the social and economic effects are concerned, recent affairs have brought us the bitter fruits of our attitude of indulgence in triviality and negligence of the essentials. California has experienced the first electricity rationing since WWII. The blackouts, cars crashing without the traffic lights, people trapped in elevators, garage doors locked dead, and most of all the crashing of computers with the messing up of bank accounts, and authors losing the complete manuscripts of upcoming books on the computer. People have suddenly waken up how much we are dependent on electricity which we have taken for granted, while concentrating on other non-essentials. Greenspan has warned that the recent California electric power crisis may lead to a national economic recession. Wake up! Complaisant dreamers! Save your life first before other secondary favorites.

The recent cold weather also hurts us by rising heating cost and deaths from lack of heating resources, which we also take for granted. It seems that we care for Marie Antonette's cake more than Parisians bread. But do not forget the price paid is the Queen's head. We pay more attention to snail darters than humans. (Actually the fish has been transplanted and thrived in other streams.) Of course we have compassion to the Earth, but life first. Stop the wistful groans for minutiae. There are a dozen more serious, pressing issues we have to deal with before it is too late, one of which is the supply of reliable, safe and plentiful energy resources which the Bush administration has announced to pursue recently. This problem will be discussed elsewhere.

It was a blunder that should never happen. But it did happen with glory, fanfare and self-gratuitous satisfaction. The arrogant leaders of the society promoted it mistakenly before; the ignorant mass follows blindly afterwards. The world is brought to the brink of disaster.

"The entire world is muddy but I only am clean; everybody is drunk but I only am sober," wrote the poet Chiü Nüan, the Father of Chinese Poetry, 2500 years ago—a perennial quote in Chinese literature. He was then asked: "If the whole world is muddy, why not dig up the dirt and stir up the waves? If everybody is drunk, why not chew up the mash and suck up the alcohol? The reply:"I understand the newly bathed cleans the robe and the newly washed dresses the hair. How could I subject my pristine body to the filthy whirlpool. I would rather go to the sparkling Miro River and live with the fishes and shrimps." Thereupon he committed suicide by drowning. Thus arose the legend of the origin of the Chinese delicacy *Tsung-Tze*, available in all *Dian-Sum* restaurants world over, originally designed as an offering to the poet by dropping in the Miro River. The food was wrapped in broad leaves tied in strings to fend off interlopers to reach the intended destination. By the way Boltzmann and Ehrenfest did commit suicide. And there are always Socrateses that will fight to death against the subtly fallacious sophists.

Appendix

A 10^9 Year Climate History and the Evolution of Humanity

a. Planetary Evolution of Mars Earth and Venus

If fossil fuels sould last all the way, passing the 300 year limit and continuing beyond 2000 years, then the sky would be fully covered with clouds and there would be no more room for adding new clouds to reflect away sunlight to cool down the earth, which would then begin to warm up. But by the law of phase transition, the heat will be used to melt the polar ice first. After it is melted through and the fossil fuel supply should continue, the earth would then warm up in earnest—the unveiling of the much heralded greenhouse warming. Eventually the ocean water would boil dry and the earth temperature would reach 600°C. The earth would become a Venus! Wow!

On the other hand if in a glacier advance of an ice age the glacier would not stop at the usual place of 40°N latitude (nearly the Mason-Dixon Line) and keep on advancing to the Equator, then the earth would become entirely frozen. The resulting ice ball would reflect away nearly all sunlight and cool it down to sub-zero temperature forever. The Earth would become a frigid Mars. Wow!

Actually planet Mars has an *eccentricity cycle* period about ten time longer than Earth and would have ice age period about ten times longer (see later) so that it would be capable of doing so. Space exploration has shown that Mars once had an ocean filled with water. Now all water has turned into ice piled up in one pole. The other pole is piled up with carbon dioxide ice.

This answers a puzzling question in planetary evolution. Whereas the three comparable *terrestrial planets* Mars, Earth and Venus were born as fraternal and almost identical triplets by mother sun in the solar system; whereas they have nearly identical *infanthood*—all began as cold solids resulting from accretion of planetisimals; whereas they all have nearly identical *childhood*—going through a complete melting by primordial radioactive heat, and then cooling down to form a thin crust with a molten interior providing occasional volcanic displays as we have now on earth with ocean formed and filled with water; their *adulthood* is completely different with destinations poles apart, from the frigid Mars to the steamy Venus. Only Earth has the extremely slim chance and good fortune to become a temperate abode to bring forth life and humans. A slight misstep in the evolutionary history could end up in either hellish worlds. Thus the probability of life outside earth is much smaller than astronomers estimated by just counting suitable planets born outside the solar system in the Milky Way galaxy. The crucial factor leading to this difference is the climate change of the planets as demonstrated in the above—a slight difference in the beginning can lead to tremendously large changes later.

b. Search for Intelligent Life Outside Earth

Even more puzzling is the emergence on earth of humans above other lives. Here is the barricade separate science and spirituality that no one has trespassed. But the border is never closed. The other primates today cannot evolve into humans because they are as usual fully specialized at the end of the blind alley of evolution. Furthermore the predecessor of Homo Sapiens cannot evolve into humans by Darwinian evolution, which is materialistic that excludes spirituality. It requires a non-materialistic evolution theory which manifested rarely in history (see

WWW.PeterFongBooks.com). Accordingly the probability of emergence of intelligent beings is much smaller than Carl Sagan has estimated in *Cosmos*. Again all this change is largely determined by climate as will be shown shortly.

I estimated the probability to be just 1 species in our galaxy. We are the only one in the Milky Way. This settles the issue of the *multiplicity of the world* which had condemned Bruno to the stake for burning. Thus interstellar communication with other intelligent beings is hopeless.

On the other hand this calculation actually confirms the existence of such beings in other galaxies, one species to each galaxy. The total number would be billions. But the radio communication would take thousands of years to go and return. We cannot wait that long and the billions of dollars to be spent on interstellar communication and searching for life beyond earth are meaningless and to be wasted in vain. Virus in crystal form can exist on Mars indefinitely. What is the use of finding such a life outside the earth?

c. Origination of Humans in 10^6 Years

Given a suitable planet, chemical evolution can explain the origin of organic compounds. Darwinian evolution can explain the emergence of life up to the apes. The first sign of emerging of humans is the appearance of the *hands*. This has been explained by the *pre-adaptation theory* that human ancestors once lived in forest and there hands for grabbing is more useful than foot for walking. In fact many monkeys today are indeed *four-handed*, and became over-specialized and trapped in forest.

Why did humans avoid becoming four-handed monkeys and become the unique species that is *two-handed-and-two-footed* and the master of the world? This is an example of the *emergent evolution* with a logic break

1997; see www.PeterFongBooks.com. Briefly, ice ages come and go; so are the forests; and so are humans moving into and out of forests. Hands are good both in and out of forest and fore limbs developed straight to hands. Feet are good outside but not inside forest, and thus rear limbs oscillate back and forth and remain in feet. Together with other influences, it may well be said that the human is the child of ice ages. An understanding of the climate of ice ages and its origin and mechanism is the pre-requisite of understanding ourselves.

d. Origin of Civilization in 10^4 Years

Not only that, humans did not stop there but further developed humanity and civilization, another spectacular *emergent evolution*. The bridge across the chasm between humanity and bestiality is *language*, which is *not materialistic, nor spiritual but totally scientific*. The problem may be highlighted by the following question: Why can apes understand English, but not speak English and cannot become humans; and yet why can parrots speak English but not understand English and cannot become humans? That shows the emergence of language is the crucial and rare turning point on the way to humanity. It is here the non-materialistic, non-Darwinian evolution theory above mentioned came in (reference *loc. cit.*).

Moreover, language is not only a *means of communication* but also a *median of memory* (such as memorizing the Lord's Prayer), just like money is not only a median of exchange but also an instrument of saving, thus arising a science of money and banking. The *memory box* in the brain is the only missing computer component of the brain to work as a computer. But so far we have not the slightest idea of what and where is that mysterious box. However, with *language memory* working as the box, the brain may be scientifically analyzed and operated as a computer, just as economics may be studied by money and banking in a respectable science, while outside materialism and spirituality.

Furthermore, with language extended from *oral* to *written*, the memory box has extended from the brain to the *books*, and the entire *library* has become an *extra-mural brain* with information storage containing all the wisdom of humans of all civilizations, which is the ultimate product of the history of evolution of life. This is yet another, and the ultimate, emergent evolution toward humanity initiated by written language (the word—eventually The Word via *Verbum* via λογοσ).

This is a *breakthrough in brain science* just as the Watson-Crick rule in molecular biology. The rest of brain activities can then be reduced to merely computer mechanisms and studied mechanically. The details will be published elsewhere. See WWW.PeterFongBooks.com for a preview.

The development of oral language is likely to take place in 10^6 years and marks the unique feature of Homo Sapiens apart from other primates—savage humans can be taught English but apes cannot. It took a long time until 10^4 years ago that written language appeared, which is then accompanied by civilization. What happened in the last 10^4 years? Again it is the climatic change—the appearance of a long temperate *interglacial* period, which is the incubator of civilization. The science of the origin of the warm interglacial will follow at the end (Appendix h).

e. Climate Influence and the Ice Ages

A crucial part in this evolution toward humanity is the *climate influence of the ice ages*, the alternating warm and cold of which provide *variation and stimulation* to prevent over-specialization to be trapped in an evolutionary blind alley but also general *stability* in evolution to prevent extinction and opportunity for drastic expansion. Humanoid emerged in recent times dominated by the ice ages; human brain enlarged three times in the last one million years and human civilization started 0.01 million years ago at the end of the last glacial retreat. Thus the ice age is not only of interest to geophysics but also to biology and humanity.

Our study of greenhouse warming is essentially based on the presence of the ice age, which is the origin of the permanent ice on earth, and is merely a sesame-seed sized tail-end story of the geological history of the earth. A closer look at the big picture of the ice-age dominated climate evolution is instructive for the problems of our main concern.

Since the discovery of ice ages more than a century ago, it has been a big mystery and more than 30 theories have been advanced. The problem has not been considered solved because new theories appeared once every other year. Most are concerned with the explanation of the frequency of the ice ages, which is once every 0.1 million years. Just about every phenomenon in the universe that has a frequency of 0.1 million years from the distant cosmic dust clouds to the bottom of the deep ocean have been invoked to explain the origin of ice ages. The two most recent ones are: (1) the changing moon orbit affecting the tides (Oak Ridge) and (2) the oscillation of the earth orbit plane in the solar system (Berkeley). All evidences are circunstantial, not even *Prima facie*, let alone smoking gun. Nearly all can be dismissed by the simple question: "If so, then why did ice ages not appear in most of the geological history before Pleistocene a mere 2 million years ago—why is it not perennial as implied by all the theories?" Failure to come up with an answer to this crucial question in any theory, several years of hard research work go down the drain for nothing.

Now, there must be a separate problem of a *catastrophic beginning* of the ice age independent of the frequency issue, but no one has paid any attention to it. Furthermore, if the theory is good enough to get the glacier advance started, it is likely to be too good to go all the way to the Equator and would make Earth a planet like Mars, which is another disaster. And why did the glacial advance always stop at the Mason-Dixon line? The problem looks like a hard to please, meticulously nit-picking slave driver.

f. The Origin of Ice Ages

One of the early and much quoted theory is that of Milankovetch, which considered the change of the *earth orbit eccentricity* of a period of 0.1 million years as the driving cause of the ice ages. The eccentricity induced temperature change of the earth can be calculated by *fluid dynamics* and the result is a few tenth of 1°C (Sellers), ten times too small for the ice ages. The theory was thus dismissed on this account. The theory also fails to account for the catastrophic beginning.

I recognized the supreme importance of the catastrophic beginning over and outside the frequency and all others issues, and proposed a complete outsider—the *continental drift* that brought the Antarctica continent to the polar position shortly before Pleistocene (the starting of the ice ages)—as the catastrophic beginning, which is needed in any theory of ice ages and seems to be a good choice because it makes snow to accumulate on earth in the polar region to form a *permanent ice* to cool down the earth.

Once there is a *permanent ice* there is ice-water *phase equilibrium* (in the iceshelves of Antarctica), in which water may change to ice to enlarge the permanent ice (just like iced orange juice put in the freezer will see the ice cubes grow in size) causing glacier advance. By the law of phase equilibrium it needs only a small temperature difference to initiate phase transition and the eccentricity induced small *temperature change* mentioned above is enough to start the process, though slowly. How fast it proceeds depends on the rate of supply of *heat*, the more heat the faster.

It so happens that the *eccentricity induced heat* generated in an ice age is of the same order of magnitude of the *latent heat* of phase transition of all ice created in the ice age, and thus is capable of starting and sustaining a glacier advance and in synchronization with the eccentricity cycle. This is an important fact that no one has recognized its significance. None of the 30 theories have given an account of this energy balance, which is the heart of any problems of this kind and should be given first order importance of consideration.

To identify the *eccentricity heat* as the one to supply the *latent heat* for the glacial ice formation requires a calculation that every college freshman can do but no experts have done it and missed the most important clue to show that the eccentricity forcing is the correct answer and dismissed it forever and engaged in a wild goose chase throughout the universe in vain. This brings out once again the snafu of mixing up *heat* and *temperature*.

To be fair I have not done that college freshman level calculation in the beginning as a prophet, but learned it the hard way backward from the solutions of my differential equations of the final mathematical theory by accident, showing how unforgivably stupid I was as a repeat of the challenge to the second law of thermodynamics. Even if I had done, it is still a long way to the end because a major question has not been answered—why the small changes of temperature and heat from eccentricity can generate ten times as large changes of them in earth climate during the ice age.

In the meantime another important lesson learned is to recognize the working between philosophy (laws of nature) and history (evolution of the world). The two are distinctively different but mutual complementary. Both are indispensable and cannot be mixed up. The catastrophic beginning is obviously a historical, evolutionary issue in the evolutionary history of the earth, not to be found in Newtonian fluid dynamics, as most researchers try to do. Missing this point, most of the 30 theories of ice ages are wasted in vain. Philosopher Kant has said history without philosophy is meaningless and philosophy without history is empty. The current academic trend of narrow specialization without a broad all-encompassing viewpoint could lead to results both meaningless and empty.

Academician M. I. Budyko in a private communication pointed out that the steady cooling of the earth temperature before Pleistocene sets the stage for a small forcing, such as the Milankovitch's, to start the glacial cycle. He is the only one else who recognized the importance of a logically unrelated catastrophic beginning to start the glacial cycles. He ascribed the cause of cooling to the steady decline of atmospheric carbon dioxide, which, as an isolated fact, is historically correct. It serves the same purpose as the continental drift of Antarctica of cooling down the earth, but the latter has the additional advantage of creating a permanent ice on earth to start the *neutral equilibrium of phase transition* that dominated world climate as we will learn in the ice age theory. For a *small* forcing such as Milankovitch's to start a *large* glacial phenomenon, he appreciated my "trigger mechanism," (which will be explained as a triode amplification in the following shortly) and wished me great progress in my subsequent research.

The last issue to confront the Milankovitch viewpoint is why the ice age is cold by a few degrees instead of tenths of a degree as fluid dynamics first predicted. The answer is of course the ice itself—the *newly* created ice makes the earth colder by a few degrees (the ice box theory). Cold was thought to be the *cause* of the ice age but actually turns out to be the *effect* of ice age, again a naive blunder of mistaking the *effect* as *cause*.

g. Ice Age Dynamics

To be explained elsewhere are the *albedo and infrared feedbacks* that work between heat and temperature to bring out glacier advance and retreat as they actually happened. They together make the climate system work like a triode amplifier in which the grid potential wave is amplified in the plate current with the same frequency (without violating the energy conservation). Thus the temperature change is magnified tenfold.

The underlying principle is the *neutral equilibrium* just mentioned by which a small change can generate a large effect like a gentle push of a cylinder on a smooth plane can make it roll on over a long distance. Neutral equilibrium is in every freshman physics textbook but no climate scientists have taken advantage of it in solving problems involving phase transition which is at the center of our consideration as it appears in the Le Chatlier principle. With this idea all problems can be worked out in a comedy of errors (including those of myself) to fool, trick and disarm the nasty slave driver, and the mystery of ice ages can be broken. The ideas are put together in a mathematical theory, the differential equations of which are solved and quantitative results checked out with experimental results [Peter Fong, Climatic Change 4, 199-206 (1982)]. The intricate interactions of cause and effect, as well as philosophy and history are explicated crystal clear. The smog that has fouled the climate theory is blown clear.

The Pleistocene ice ages represent a period that exhibits strict regularities in large magnitude variations in the wild frontier of earth climate and thus provides the precise testing ground to screen all possible theories. The winner of the contest deserves to be the guide post of future studies, such as the anthropogenic greenhouse warming.

This is the only mathematically calculable and numerically displayed physical theory without any adjustable parameters and with adequate physical explanation of the intricate insights of the convoluted mystery of the ice age. The only challenge to the theory I can think of is the carbon dioxide issue—is it the cause or effect of the ice age? Both are logically viable (details omitted). If the gas is the cause then the theory must be basically revamped. If the effect, then the theory needs merely a little touch up without major alternation.

It is again not a philosophical (rational) but a historical (circumstantial) issue. The evidence is the *Vostok ice core* determination of *carbon dioxide* and *methane* (both greenhouse gases) contents in ancient ocean over the ice age. The results show the variation of contents of the two gases over the ice age are the same, indicating they are results of the same cause—the ice age. Thus the carbon dioxide variation on earth is the *effect* of the ice age. There is no reason that the presence of the two gases in the air is so related because they are two logically independent gases (gas in air and water are related by equilibrium). Thus carbon dioxide cannot be the *cause* of ice age. (If it is the cause then what is the origin of this cause? The buck is passed but the problem is not solved.) As will be shown shortly carbon dioxide played a minor role in climate history.

Carbon dioxide was not included in the above mentioned mathematical theory of 1982, but can be incorporated into it without changing the main outlook of the conclusions because it is an effect, not the cause.

Over all carbon dioxide is not a principal actor in the present climate drama. In the Silurian Period carbon dioxide is twice the present day value but the world climate had an historically low temperature comparable to the present time and there existed glaciations. In conclusion ice is more important than carbon dioxide in climate determination; the latter is an accompanying consequence within a million year time scale. In fact in the history of the Earth carbon dioxide content had reached a maximum of 16 times the present value in early Peleozoic Era, due to extensive erosion resulting in great sedimentation,at about the time of the second temperature minimum in history corresponding to another era of glaciation 600 million years ago. In the main text this big picture is reduced to a minute segment in recent times and then magnified tremendously to show the details.

Incidentally the catastrophic beginning introduces the idea that historical facts are not necessarily logically connected (but are accidentally brought together by chance) which is at the basis of the currently popular *emergent evolution theory* which has the advantage of transcending the materialistic Darwinian theory, a point related to the origination of humans. The synthesis of the heavy elements of carbon-12 and beyond in stellar evolution and the creation of the present world including all of us depends sorely on the existence of a resonance level of the nucleus carbon-12 at the energy of 7.657 Mev (experimentally verified at Oak Ridge) which is exclusively determined by the structure of nuclear physics with no logical intention of creating the present world. But without that resonance level at 7.657 Mev, none of us will be here today to discuss any scientific problem.

In any research study it is important to separate the true cause from the irrelevant ones. The difficulty is that it is not merely a matter of logical certification but the intrusion of many historical and other accidents. Also important in proving a case is to distinguish and evaluate accordingly the circumstantial, *prima facie* and smoking-gun evidences as the lawyers do so that the weight of the evidence can be assigned accordingly.

h. The Warm Interglacial and the Incubator of Civilization

With the ice age theory clearly established, we have learned the dominant factor controlling the Pleistocene and Holocene climate. Thus this may be used as the basic framework to study the current greenhouse warming as a continuation of the geological history of the last ice age.

However, as we are now in the *interglacial*, this concern exposes a new twist that the warm period of interglacial is in reality much longer than that expected from the raw eccentricity theory, a new basic as well as a prerequisite problem.

Our basic framework is one of ice-water phase equilibrium, which is in a *neutral equilibrium* tending to stabilize the earth temperature at the transition temperature of the phase change. The equilibrium is perturbed now, besides by the earth orbit eccentricity but in addition by the 300 years of anthropogenic carbon dioxide, which is a tiny fraction of the carbon budget of the earth surface and is a tiny tail-end extension of the story and will not change the broad picture. The ice-water neutral equilibrium will be expected to continue to maintain a constant climate of the earth during the interglacial including the 300 years, with the permanent ice of

Antarctica acting as a giant *thermostat* to regulate the random climate fluctuations.

However, in the period of the interglacial the polar ice is too far away from most water of the earth to interact effectively and therefore the polar ice-water equilibrium is interloped by another actor, the *water-vapor equilibrium from a phase transition* from moisture to water droplets *in the cloud*, which is also a neutral equilibrium (more later). Thus the clouds form an *intermediate thermostat* in the equilibrium process. In this way we come to the idea that the interglacial climate is controlled by the *three phase equilibrium of water*—ice, water and vapor in different geographic locations during this period as stated in the main text as a principal point.

Following this basic idea the *history of the interglacial* will be as follows: Note that all neutral equilibrium will stop when the thermostat becomes dysfunction. The intermediate thermostat clouds becomes dysfunction when the sky is fully covered with clouds if fossil fuels last for 2000 years more. After which the polar ice will begin to melt and that thermostat will become dysfunction when the ice is melt through if fossil fuels continues forever. Then would be the end of ice ages and Pleistocene on the earth which would return to the *Pliocene* climate and become warmer. In reality this history is valid only up to 300 years from now.

The interjection of a new actor of water-vapor neutral equilibrium at the last minute of the theory inevitably arouses the suspicion of a conspiracy of collusion. A solemn, no-nonsense statement is necessary to clear the point—to show that it is not only scientifically necessary but also answers the most intriguing question in all sciences: why did human civilization originate in the present interglacial (8000 years ago shortly after the receding of glaciers 11,000 years ago).

During the glacial period the earth was cold and the sky was clear without cloud—all cumulus clouds were precipitated as snow (the present cloudy winter sky is no excuse). In the mathematical theory above mentioned the sky was assumed cloudless to begin with and throughout to bring out the main features of the ice ages, in spite of the fact that it is excepted in the interglacial periods, which were short and thus justified to skip as a good approximation. The present day climate is in the interglacial and for the current greenhouse warming problem the special situation in the interglacial must be investigated in detail, not smoothed over as in the million year history.

By the way the omission of the cloud as a variable in the above mentioned theory, as well as the omission of carbon dioxide, both being significant considerations of the ice age problem, have not been justified until now. The theory of 1982 *per se* was not complete for the two omissions and others, arousing strong controversies and was published on probation (see the Editor's Note), but is completed now. It was done at a time without necessary information to guide from, but ventured out by a *Fermian insight*, which I learned from Professor Enrico Fermi, who can see through the mist of complexity to grab the critical point and clear the briar patch mercilessly to arrive at a simple, if not complete, solution, (with a quality control like the spaceship systems by cross-checking the cause-effct network even there is still unknowns), that will stand the *test of times*, which is more important than the test of *experiments* and *theories*.

At the end of the last glacial retreat 11,000 years ago, the sky remains clear without cloud but the melting of ice has stopped, evidenced by the presence of Ross iceshelf and others in Antarctica today. Instead of turning cool to start a new ice age as expected from the eccentricity cycle, we had *a long warm interglacial* lasting to this day. This is unexpected and has not been explicated. The reason may now be explained as follows:

Emiliani et al have established by deep-sea cores [Science **189**, 1083 (1973)] that there existed an episode of *rapid* ice melting and sea-level rise at about 11,600 years ago, coinciding with the age of the *Plato flood*. (There are many other ancient flood legends including the Bible.) The reason of the *rapid* melting has been attributed to the breakup of the remaining continental icesheet in the Northern Hemisphere with the fragments sliding into the ocean quickly, which caused the ocean level to rise quickly, resulting in the legendary flood which refilled the Mediterranean Sea that was drained dry during the ice age, inundating the primitive agricultural communities all of a sudden. That time may well be the beginning of the warm interglacial.

Since the melting is *rapid*, the glacial cycle stopped abruptly and lagged behind the eccentricity cycle, which has not finished the warming cycle and the eccentricity generated heat was still impinging on the earth but was too far to reach the small and distant Ross and other iceshelves in Antarctica. Thus it was interloped by the ocean to evaporate and generate clouds, something first appearing in the sky after the ice age.

It took 4000 years to generate enough cloud to cover the earth entirely with clouds (eccentricity forcing being comparable to greenhouse forcing). By that time the eccentricity cycle has turned to the cool part and it took another 2000 years to reduce the cloud cover to one-half to the present day coverage, that is where we are in today, and will take another 2000 years for the cloud to be reduced to zero, with a clear sky to start the next ice age.   (Here we take advantage of our result of greenhouse studies—2000 years for cloud change—to study other climate problems.)

Thus there is a gap of around 13600 years between the end of the last ice age and the start of the next—a fairly long period designated as the interglacial with a very stable temperature determined by the *phase equilibrium of water and vapor in clouds* (not ice). In this period the climate is regulated by the cloud thermostat (not the polar ice thermostat). This explains the longer then expected interglacial. Turning things around, the very existence of the interglacial is the ultimate cosmic proof of the absence of global warming today because of the cloud thermostat. From a logical point of view the problem should start from the interglacial then to the global warming but historically we happened to learn and solve the global warming first then go back to the interglacial. The same is true that we happened to solve the global warming and ice age first then go back to planetary evolution contrary to the time sequence order.

This also explains why the next ice age has not started today and will start only about 2000 years later, contrary to the superficial prediction of the eccentricity cycle. Factually the glacial cycle lagged behind the eccentricity cycle by 30° in phase corresponding to an interglacial of 10^4 years, a nagging theoretical problem, but a good news for the world community for delaying the onset of the next ice age. This theory of the long interglacial solves the theoretical problem and affirms the good news. Moreover, for our purpose, the cloud thermostat is the basis of our view that there will be no global warming in the next 2000 years.

A bonus is to answer the question why human civilization started in the past few thousand years. It is because a reasonably long time with good climate is needed to accomplish the job—the development of agricultural communities, the complex human interactions, the written language and the accumulaton of wisdom to spwan civilization. Without such a long period, the raw eccentricity driving force would change warm to cold "instantly" without a long incubation period to do so. As a matter of fact, reasonably long (not instant) interglacial periods did exist in past Pleistocene history. Why did civilization not begin in the earlier interglacials? The answer is not to be found in a philosophical inquiry but in a historical accident. We *happened* to be in the *first* interglacial that civilization has happened.

Schneider has advocated the peril of the next ice age before promoting the hazard of global warming. When pressed for the contradiction, he answered that both theories are correct, which is indeed true from a purely logical point of view. The only point laymen did not appreciate is the timing of the events. Thus we have a right to worry about the peril of the next ice age and to find ways to salvage our future decedents 2000 years later. Greenhouse warming heat is just right to compensate off the cooling of the ice age and would be a blessing. Unfortunately fossil fuels lasted only for 300 years. On the other hand if humans are to survive and thrive, there must be increased use of energy, and any energy after use becomes environmental heat, which may eventually be much greater than that generated by all fossil fuels and therefore may be used to stop the glacial advance. The only source of energy for that purpose is nuclear energy, which, alas, is another anathema of the prevailing popular view, but will be the bitter medicine everybody must swallow in order to survive. Nuclear energy is another important issue that will be discussed elsewhere.

Hypothetically if without the polar ice today, the current cool earth would return to the average warm earth of the past history and the future history would be completely different. If we start a climate theory with today's conditions but omitting the permanent polar ice, then the truncated initial conditions so presented would not be self consistent logically and therefore all mathematical derivations that follow would be complete nonsense—garbage in and garbage out as the computer addicts well know—no matter how perfect is the logical structure of the theory and how high powered is the computer used. The mistake made on the initial condition is the most serious blunder but is one that happens often.

Besides the *initial condition* the other important element is the *dynamics*. The popular idea of greenhouse warming and its related policy measures are limited to a short period of the life of fossil fuels in the interglacial period in which the climate determining factor, i.e., the *three-phase equilibrium of water*, overrides *fluid dynamics* which merely determines the *weather*. The current study of global warming based on *fluid dynamics* is thus as good as weather prediction which is accurate for a week and useful for a year. For climate study of 300 years at the present time the ice age theory controlling the glacial and interglacial periods is a better guide than relying on the super-computer, and neutral equilibrium of ice, water and vapor sets the tune of most changes.

i. El Niño and La Niña

Having explained the origin of the long interglacial, it is the opportune time to return to the issues of El Niño and La Niña mentioned earlier in their influence to recent climate. It is a problem similar, but in a smaller scale, to the ice age; it concerns the alternative changes of equatorial ocean surface temperature and their effect on world climate. A major difference is that the cold or warm interludes do not have a constant period (2 to 3 years) and appears haphazardly, which are different from the ice ages. It has been recognized as a problem of interaction of the ocean with the atmosphere. Philander has studied it by the interacting equations of the ocean and atmosphere systems with success but acknowledged remaining problems.

The *Fermian insight* leads the way as follows: First to recognize is that it is a statistical, not dynamical, problem. Energy balance is more important than detailed kinematical explications. The current interglacial climate is in a state of three phase water equilibrium. Any changes are fluctuations away from that equilibrium. Some can be cyclic as oscillations. The study of ice age as a cyclic system provides a guide. The absence of a sharp period and a catastrophic beginning indicate it is not a *forced* oscillation, but a *free* oscillation. Thus there is no need to search the universe for the outside cause. Greenhouse effect, a late comer, is not.

A large *free oscillation* must be traced to an instability, which can be magnified by a *positive feedback* and then be stabilized by a *negative feedback* to return to the starting point, ready for the next cycle of oscillation (just like the erratic behavior of the stock market not expected in a stable economy; there is also possibilities of *forced oscillations* caused by external factors, such as wars and the increase of population, capital and knowledge of science and technology). If the initial fluctuation is an increase of temperature, it manifests in an El Niño; if decrease, a La Niña.

One positive feedback is readily available: A fluctuation of temperature increase of the ocean surface may increase the evaporation of the ocean and thus the water vapor density, whose greenhouse heating may increase the water temperature again in a diverging positive feedback. Once the temperature is high enough, a ready-made negative feedback, the *infrared feedback*, is available to pull it down to normal, which is accelerated by the same positive feedback now that the change is a negative one. This can happen anywhere in the ocean (leading to tropic hurricanes).

However, in *equatorial Pacific ocean* the prevailing wind, before the cooling down, blows a large body of warm water eastward toward the coast of Peru and accumulated there, as long as the wind is blowing, to form a large body of warm water for 2 to 3 years. This is the origin of El Niño as the conventional wisdom has it. After the 2 to 3 years, negative feedback will start to reverse the course and to return to normal.

Before the gross balance of the equilibrium is re-established, during the 2 to 3 years of El Niño, the climate state is upset, detailed balance lost; it is in a turbulent state. Climate change is destined to happen drastically.

Subtracting the Maxwell-Boltzmann distribution from that of the turbulent state, we have a distribution representing the added turbulence. Its *vector velocity field* may be decomposed into three components: 1, a *gradient field* that will turn into a directional wind; 2, a *curl field* that will turn into an eddy current in cyclone wind; 3, a random current field that will turn into heat. The eddy currents tend to evolve into a smaller vortex with higher circular velocity, due to conservation of angula momentum, resulting in *hurricane or tornado*. The directional currents, by linear momentum conservation, lead the hurricane moving slowly for hundreds of miles in the ocean. The origin of the vertices is the Coriolis force due to the earth rotation. It also sucks in large volume of moisture resulting in *torrential flood*. The depletion of moisture in outside areas results in *draught*. In an equilibrium system the conservation laws hold, which explain the warm and flood of the south must be compensated off by the cold and draught further north. This contradictory behavior of flood and draught at the same time is typical of El Niño, not easily explanable by laminar flow in fluid dynamic but more likely by turbulence resulting from the breakup of an unstable equilibrium (cf. the *avanlache theory*, the potential energy of which is now played by the latent heat of condensation). The energy balance in earth temperature changes in recent years as related to ocean changes in El Niño and La Niña has been quantitatively discussed before. China has suffered from the double curse of south flood and north draught in the same year once every few years. It would be interesting to study the correlation with the El Niño in the Pacific ocean. Eventually the gross balance will re-establish the equilibrium state and the climate will return to normal. This explains the El Niño and La Niña phenomena as aberrations of an otherwise tranquil climate in the interglacial.

The El Niño problem is a miniature of the ice age and exhibits general principles applicable to global, complex problems that merit mentioning. Since time immemorial humans have searched for cause-effect relations beyond mythology. But the concept of *cause* is vague. Aristotle has distinguished four kinds of causes. Buddhism, two: the deterministic cause (yin) and circumstantial cause (yuan), excluding all others; thus it abolished all Hindu gods who have become redundant, as Newton has done later. Buddhism is a rational religion, forecasting modern science. These two concepts appear in the Newtonian system of science as *laws of force and motion* (differential equations), and *initial conditions* respectively. With these two the mathematical theory of differential equations provides a unique, complete, and exhaustive solution and the problem is solved once and for all without room for further dispute. Mythology is finally replaced by science—the one and only science.

The differential equations represent the philosophical aspect of the problem, which is constant, universal and timeless. The initial conditions represent the historical aspect, which is variable, arbitrary and irrational. The inter-weaving of the two forms the tapestry of historical events, which exhibit both contradictory features—order and discontent, justice and compassion, fanaticism and tolerance, serious and humor, passion and abandon, glory and humility. Thus is the *emergence* of *humanity* in history. The understanding of these requires a unique foresight to unify the contradiction of philosophy and history which is stated succinctly by Kant's aphorism quoted earlier and Sma Chien's dictum well known in Sinology. Short of that, things can become meaningless or empty. The notion that environmental cleaning can usher global warming is totally meaningless and empty—a thought that is intellectually bankrupt. In addtion to the *Poverty of Philisophy*, we may also have the poverty of history.

Common mistakes are the following: 1. Mixed-up of cause and effect. In an historical event made up of many episodes, the effect of the first episode is naturally the cause of the second episode. In an oscillatory event the cause and the effect of an episode are exactly the same (so that the episode can repeat itself exactly in oscillations). On a *prima facie* look you cannot decide it as cause or effect without an historical perspective, and may mistake the effect as cause as in the ice age problem.

2. Neglect the initial condition which can be arbitrary and irrational to the problem under consideration, such as in the ice age. In the Berkeley theory of ice ages based on orbit-plane oscillations the unavoidable initial condition is attributed to the cosmic dust cloud—a buck-passing maneuver.

3. Mis-gauge the time scale of the events—short time problems are largely determined by the *initial* conditions (through dynamics) whereas long term problems by the *final* conditions (through equilibrium)—a point Fermi has often emphasized. Thus in the El Niño problem the energy balance outweighs the kinematics. In these ways mistakes can be made and

corrections can be done. In fact these are all the same in studies of all sciences and humanities.

So is the way to look at the rise and fall of the stock market.

So is the way to look at the rise and fall of an empire. By the way one cause for the fall of an empire is the rise of the Marie Antonnette syndrome. Eighteen hundred years ago a super El Nino led to 6 inches down-pour of a flood in Peru. The Mochica Indians pleaded their god by human sacrifices of their own kind. The deluge continued and so the human sacrifices, a well-developed pre-Columbian civilization was wiped out. Mythology is no help in dealing with natural phenomena. The greenhoue warming has had its first shimmer bearly seen, which may well be a phantom, but a massive sacrificial rite has already began, which may lead to an economic depression as Greenspan has already warned. As *Mother Goose* has it, the loss of a horse shoe may lead to the loss of an empire, the economic loss may lead to the collapse of *Pax Americana*.

Postscript: Concluding Remarks

This is an indictment, trial and sentence of the fraud case of global greenhouse warming. The exhibits include the data of precipitation, cloud coverage, temperature, flood, draught, hurricane, El Nino, La Nina,...; the witnesses include Einstein, Maxwell, Boltzmann, Fermi, Wigner, Yang, Lee, Prigogine,...; the verdict is guilty. The legal foundation is statistical mechanics including the second law of thermodynamics and the Le Chatlier principle. The principal arguments are contained in the main text.

The Appendix after that contains a number of related topics upporting the main arguments. They include the planetary evolution, the evolution of the earth, the ice ages, the emergence of humans and civilization, the El Niño and so on, all important and interesting topics by themselves. They actually constitute a second essay *A Billion Years Climate History and the Evolution of Humanity*, which can stand by itself.

The logic of these topics and the main arguments of greenhouse warming are tightly knitted into a seamless web, like Jesus' robe that was woven from top throughout and could not be rent (John: 19, 23). Anyone challenging the verdict will have to come up with an entirely new cosmic view, to uphold his intellectual integrity, not just to rend a piece of the robe to play clownish antics as the *Sorcerer's Apprentice* in *Fantasia*.

The controversy of greenhouse warming will be as ephemeral as McCarthyism, a blood boiling issue that will vanish without a trace in a short time once the fraud is apparent. The bludgeon orgy fight today will soon be forgotten and the main text will not be read by any one any more after that. But the Appendix will stand for 100 years. After all, greenhouse warming is merely a tail-end story of a spectacular epic drama. How did it unfold is a fascinating question.

The *neutral equilibrium*, the ugly ducking in freshman physics, turns out to be an exquisite swan. Like a shepherding angel, it guided the world through a perilous journey, avoiding pitfalls and booby traps, over planetary evolution, ice ages, interglacials, human evolution and reaching the present grand apex. It provides stability while allowing progress, without the disasters of turmoil and stagnation. The least we can afford now is a frivolous cavort to tumble from the marvelous apex to the irretrievable precipice.

LATENT HEAT OF MELTING AND ITS IMPORTANCE
FOR GLACIATION CYCLES*

PETER FONG

Physics Department, Emory University, Atlanta, Georgia 30322, U.S.A.

Abstract. A dynamical energy balance model is developed including both latent heat and sensible heat exchanges. It is applied to reconstructing the history of the changes of the icesheet mass and the ocean surface temperature over an ice age. The zero-dimensional model is extended to include three-dimensional information of the icesheets by assuming a specific geometric shape of the icesheets. The ice-albedo feedback can then be calculated and, at the same time, the cryosphere interaction is introduced into the climate model. The advancing of the glaciers and the cooling of the oceans in a glacial period can be accounted for by the differential equations of the dynamic system if an external perturbation in the form of any energy deficit of 0.13% of the insolation is imposed. The earth orbital changes generate a heat deficit of this magnitude due to the change of the eccentricity and have the same periodicity of 100 000 years as the major glacial cycles. Therefore they could well be the origin of the Pleistocene ice ages.

The objectives of this short paper are to present the basic ideas of a simplified dynamical theory to account for the general features of the long term climate changes of an ice age and to note the implication of the theory on the origin of ice ages. The detailed elaborations will be reported later.

In recent years the planetary origin of ice ages has received great attention. The paleoclimate record established by the study of deep-sea piston cores has been compared with the planetary changes of the earth motion including the variations of the eccentricity, the obliquity of the ecliptic and the longitude of the perihelion with frequencies of 100 000, 41 000, and 22 000 yrs respectively. Good correlation of the obliquity and precession cycles with the minor peaks in the geological record has been obtained [1, 2]. This correlation is supported theoretically by climate modeling studies [3–7]. However, the correlation of the major peaks in the geological record that exhibit a 100 000 yr periodicity with the variations of the eccentricity has presented difficulties. The variations of the insolation due to those of the eccentricity, according to Berger[8], ranges merely from −0.17% to +0.014%. Making use of the Sellers climate model[9], Berger [10] found the corresponding change of the world average temperature to be only a few tenths of a degree, a factor of ten too small—according to the CLIMAP Project study [11] the average

---

* Editor's Note: This note generated strong, but mixed, reactions from three referees. Its conclusions should thus be weighed carefully. It is published despite the cautionary reviews in order to spur debate on the large remaining uncertainties over the causes of Pleistocene glacial cycles.

Climatic Change 4 (1982) 199–206. 0165–0009/82/0042–0199$00.80.
(with kind permission of Springer Science and Business Media.)

ocean surface temperature dropped 2.3 °C 18 000 yrs ago during the height of the last glacial period. Most climate models would lead to results similar to that of Sellers'. Other alternatives have been explored in an attempt to explain the 100 000 yr major cycle including icesheet models [12—14] and oscillation models [15, 16].

The main features of the climate change over an ice age are the advance and retreat of the icesheets, and the lowering and rising of the ocean surface average temperature $T(t)$, t being the time, which determines the world average temperature. A theory that accounts for the dynamic history of these two features is what we want for an understanding of the ice age problem. Many current zonally-averaged climate models [17—22] are equilibirium models and thus are not suitable for the treatment of the advance and retreat of the polar icesheets. The dynamic general circulation models cannot be applied to problems on such a long time scale (10^5 yrs). To bring out the essential points, we shall compare the dynamical theory to be developed with zonal equilibrium models as we proceed. Many points are equally applicable to general circulation models.

The change of the ocean surface temperature in response to external forcing by itself can be dealt with by zonally averaged or even global equilibrium models. Given a heat deficit (or excess) manifested in a change of the sensible heat D_s cal yr^{-1}, the change of the temperature $T(t)$ can be calculated. However, this approach cannot be applied to deal with the ice age problem because the advance and retreat of the icesheets indicate the existence of another part of the heat deficit (or excess) manifested in latent heat D_L cal yr^{-1}, which is not included in most equilibrium theories (an exception is Saltzman's model [23]).

In zonally averaged equilibrium models, the emphasis is placed on the zonal temperature. $T(\ell)$, ℓ being the latitude. In zones where $T(\ell)$ is below a certain point near 0 °C, the surface is considered ice covered — ice-albedo feedback can then be calculated. The latent heat involved in forming the ice is not included as an integral part of the equilibrium theory. The total amount of polar ice $W(t)$ gm is an undetermined quantity. The icesheet is two-dimensional (sufficient for the ice-albedo calculation); the third dimension, the height, of the icesheet is undetermined (thin blanket approximation). The volume and height of the icesheet are important quantities for our purposes. Without a three-dimensional icesheet cryosphere dynamics and its interaction with the atmosphere are missing. These are the elements we want to incorporate into our theory.

Our objectives can be defined mathematically as an attempt to determine the two quantities $W(t)$ and $T(t)$ as functions of time, which describe the advance and retreat of the icesheets and the falling and rising of the temperature that characterize an ice age. The differential equation that governs $W(t)$ is

$$\frac{dW(t)}{dt} = \frac{D_L(t)}{L_f + C_w T(t) - C_i T_i} \quad \text{gm yr}^{-1} \tag{1}$$

where $L_f = 80$ cal gm^{-1}, C_w and C_i are specific heats of water and ice, T and T_i are the average surface temperatures of the oceans and icesheets measured in degrees Centigrade (°C). Since $C_i = 0.55$ cal gm^{-1}, the term $C_i T_i$ is small and so is neglected in the first approximation. $C_w T(t)$ is retained because of the coupling of $T(t)$ with the following equation.

The differential equation that governs $T(t)$ is

$$\frac{dT(t)}{dt} = - \frac{D_s(t)}{C_m} \text{ °C yr}^{-1} \tag{2}$$

where C_m is the heat capacity of the ocean mixed layer, assumed to be 200 m thick ($C_m = 7.22 \times 10^{22}$ cal °C^{-1}). Here we assume the mixed layer is always mixed to uniform temperature so that sensible heat can be extracted (or added) in response to atmospheric temperature changes, whereas the deep ocean remains unchanged. In a time scale as long as an ice age we may wonder if heat may be extracted from the deep ocean. In that case the C_m in Equation (2) should be greater than the heat capacity of the mixed layer. On the other hand the deep ocean in an ice age may become warmer due to the change of salinity of the ocean water which changes the pattern of vertical circulation of the ocean [24]. In the absence of a detailed hydrodynamic and thermodynamic study of the hydrosphere over an ice age, we assume the present condition prevails in an ice age as a first approximation which is reasonable because the deep ocean is always near the freezing point and is well insulated.

Our strategy to determine D_L and D_s is to find their ratio and sum. The ratio

$$R = D_s/D_L \tag{3}$$

is a function of time. To a first approximation we replace this variable by its average value over a period of glacier advance, which can be determined on an empirical basis. We know in a period of glacier advance a total amount of ice of 3.3×10^{22} gm is added to the ice-sheets (from the lowering of the ocean level of about 91 m) corresponding to an integrated D_L of 3.1×10^{24} cal (95 cal per gram of ice). At the same time the temperature $T(t)$ is lowered by 2.3 °C [11] corresponding to an integrated D_s of 1.7×10^{23} cal. The average value of R is thus

$$\bar{R} = 0.053. \tag{4}$$

The sum of D_s and D_L is the total heat deficit (or excess) D which is made up of whatever contributions from external forcing D_e plus the amounts from the positive feedbacks D_p minus the amounts from the negative feedbacks D_n, all being in the unit of cal yr^{-1}. As a start, we limit ourselves to the ice-albedo positive feedback for D_p and infrared negative feedback for D_n. The equation

$$D_s + D_L = D = D_e + D_p - D_n \tag{5}$$

of global energy balance is the basis of a zero-dimensional climate model.

For the calculation of ice-albedo feedback D_p we need two-dimensional information on the horizontal extent of the icesheet; three-dimensional information will also be needed later. Therefore we have to extend the zero-dimensional model to cover three dimensions. This is accomplished by making use of the empirical fact that the polar icesheet assumes a specific geometric shape. Once a given shape is specified, a definite value of $W(t)$ corresponds to a definite three-dimensional configuration of the icesheet and our needs can be fulfilled. For simplicity and as a first approximation, we assume the shape to be a circular cone of a radius to height ratio of 1000 based on the icesheet configuration of 18 000 yrs ago [11]. A parabolic icesheet profile was derived by Weertman [25] assuming ice flows as a perfectly plastic solid in a two-dimensional icesheet. Our icesheets are three-dimensional with a cylindrical symmetry for which the parabolic profile would not apply. Also, the parabolic profile does not include the effects of elasto-plastic and visco-plastic behaviors. Better geometry will be developed later and can be accommodated easily.

For the earth system in an ice age there are two icesheets in the two hemispheres. Again for simplicity we assume the two to be identical. Thus

$$W(t) = 0.917 \times 2 \times \tfrac{1}{3} \ \pi r^2 \ \frac{r}{1000} \tag{6}$$

where 0.917 gm cc^{-1} is the density of ice and r is the average radius of the two icesheets, also a function of time. Again better approximations can easily be substituted.

Equation (6) relates $W(t)$ to $r(t)$ and thus to $2\pi r^2$, the area $A(t)$ of ice-covered surface needed for the calculation of ice-albedo feedback. Thus

$$D_p(t) = \int_{A_0}^{A(t)} (\text{solar constant}) \ \Delta(\text{albedo}) \ \frac{\cos \ell}{\pi} \ dA(t) \tag{7}$$

where Δ(albedo) is the change of the planetary albedo when land becomes covered with ice, the value of which is taken to be 30%. This value is in agreement with the change of albedo values used by Budyko [18]. The solar constant is taken to be 2 cal cm^{-2} min^{-1}. The integral over the latitude ℓ may be approximated by using the average latitude of the newly ice-covered areas ($61°$ N for the Northern Hemisphere and $59°$ S for the Southern Hemisphere [11]).

The infrared negative feedback D_n is calculated by making use of the sensibility parameter β given by Cess [26] ($\beta = 145 \ °C$).

$$D_n(t) = \int -0.7 I \frac{1}{\beta} \cdot \frac{dT}{dt} \ dt \tag{8}$$

where I is the annual insolation, 1.30×10^{24} cal yr^{-1} and the factor 0.7 of I is the fraction of the heat income compensated for by infrared radiation (assumed fixed).

The above set of coupled equations (1)–(8) now form a complete, deterministic dynamic system. Once the initial conditions $W(0)$ and $T(0)$, and the external perturbation $D_e(t)$ are given, the equations may be integrated numerically and $W(t)$ and $T(t)$ may be predicted for any furture time t. The climatic history of an ice age may thus be worked out theoretically.

Before discussing the numerically results a point on the principle may be raised. In the zonal equilibrium model the emphasis is placed on the zonal temperature distribution

$T(\ell)$. The latitudinal dependence of the temperature is needed to determine the icesheet front (for the calculation of the ice-albedo feedback). For this purpose we need to work out the difficult problem of zonal heat transfer. In our approach this problem is circumvented. The question is what physical substitute is involved in our case that determines the icesheet front.

The icesheet front, in our case, is determined by the icesheet mass $W(T)$, which in turn is determined by energy balance, and by the icesheet shape. The shape is determined by the cryosphere dynamics including glacier flow down the icesheet slope. This is an area overlooked in the climate modeling theories. In the zonal equilibrium model the ice at the icesheet front is formed by local precipitation and is determined by local meteorology. In our approach the ice at the icesheet front comes largely from the glacier flow down the icesheet slope -- the icesheet must assume the specified shape which is determined by the glacier flow. It is not determined by local meteorology; in fact it is this flow that largely determines the local meteorology. In our approach we introduced cryosphere dynamics into the climate system through the assumption of the shape. Our physical basis is thus more realistic. In comparison the conventional thin-blanket icesheet implies no glacier

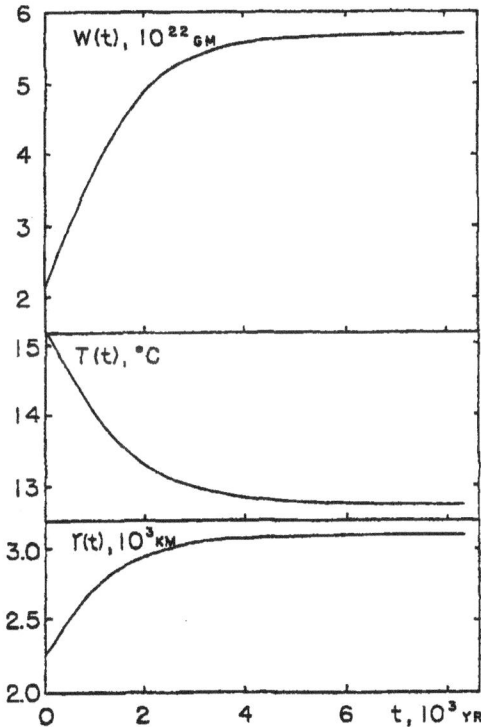

Fig. 1. The evolution of the ice age glacial system predicted by the dynamical theory. $W(t)$ is the total mass of the icecaps. $T(t)$ is the average ocean surface temperature. $r(t)$ is the average radius of the icecaps. All are functions of time t.

flow. The arduous treatment of local meteorology is rendered irrelevant when glacier flow as an important element of the climate system is included.

In carrying out the numerical integration of the above set of equations we find that the empirical facts of the last glacier advance, i.e., an increase of ice mass of 3.3×10^{22} gm and a lowering of the ocean surface temperature of $2.3\,°C$, may be accounted for by the present theory by an external perturbation of a constant heat deficit D_e of 0.13% of the insolation. With this $D_e(t)$ switched on and with the use of the initial conditions representing the climate of the last interglacial period, i.e., $W(0) = 2.2 \times 10^{22}$ gm and $T(0) = 15.2\,°C$, we integrate the equations numerically and determine $W(t)$, $T(t)$ and $r(t)$ as functions of time. The results are shown in Figure 1. The asymptotic values of $W(\infty)$, $T(\infty)$, and $r(\infty)$ agree with their empirical values of the last glacial period (Wisconsin). The agreement of all three shows the consistency of the theory but does not represent a proof of the theory because we have introduced assumptions relating the three and thus only one of them is independent. It is made to agree with the experimental value by assuming a value of $D_e(t)$ as an adjustible parameter.

The theory being dynamic, we can predict not only the asymptotic values but also their dynamic history, which is the essence of the whole effort. Additional results may be compared with observed values. The history of the system is shown by the curves in Figure 1. The results show that it took a few thousand years to complete the process of glacier advance. For one thing it makes possible to compare the rate of glacier advance with the theory. Goldthwait [27] determined a speed of glacier advance of about 100 m yr^{-1} for the late Wisconsin period across Ohio. The theoretical value from the derivative of the $r(t)$ curve at the corresponding point is 70 m yr^{-1}. This general agreement does represent an independent proof of the theory because no new parameter is arbitrarily assumed. Additional verification of the dynamical aspects of the theory is desired.

The desired perturbation, 0.13% of the insolation, for generating an ice age is of the same magnitude as that resulting from the eccentricity changes of the orbit. This supports the view that the eccentricity changes are the origin of the ice ages. The major cycle of 100 000 yrs of the geological record may thus be attributed to the 100 000 yr cycle of the eccentricity changes. The discrepancy by a factor of ten of the previous models is now overcome. There is still the problem to reconcile the previous models with the present results.

The previous models are static models, developed from atmospheric considerations, not intended to deal with the ice age problem. The equilibrium value would be reached in a relaxation time of the order of 10^2 days. We are not only concerned with atmospheric dynamics but also with the cryosphere dynamics and the hydrosphere dynamics and their interactions. We are concerned with a dynamic process lasting thousands of years. Although we conclude a temperature change ten times greater, it is meaningless to say that our model is ten times more sensitive because the change would be realized over thousands of years, not hundreds of days as in the equilibrium model. In terms of the rate of change of the temperature, our model is actually less sensitive. The concept of sensitivity of the static models simply has no meaning in a dynamical theory and therefore the factor of ten difference presents no conceptual difficulty. Our theory does bring out the long-term instability due to the cryosphere interaction.

The conventional models, like ours, do include the ice-albedo feedback. The original heat deficit induces a change of the temperature distribution, and thus an expansion of the ice-covered area, and thus an induced heat deficit, which will repeat the same in another cycle. The total effect is thus the sum of a geometric series. However, the conventional models do not include, as we do, the cryosphere interaction — the glacier flow of ice down the icesheet slope. This puts more ice on the icesheet front than it would receive from local precipitation alone because the glacier flow represents the accumulation at the icesheet front of precipitation originated higher up on the icesheet in addition to local precipitation. The ice-albedo feedback effect is thus increased. Even though the increase is of the same order of magnitude and thus does not represent a drastic departure from the conventional theories, the summation over the geometric series amplifies this difference. As a matter of fact in our case the geometric series is almost divergent. Thus the results of the two approaches differ by a factor as large as ten. The spectacular phenomenon of glacier advance is induced by a very small external perturbation because of the very delicate nature of the dynamic system.

This great instability of the earth climate system (atmosphere-cryosphere-hydrosphere) is an important point to be considered in connection with the future climate changes originated from anthropogenic perturbations, such as the carbon dioxide generation. In that case the temperature increase might be greater than that predicted by current models but the effect will not be immediately felt. It will manifest itself in hundreds of years. In the meantime the icecaps would melt, refrigerating the earth to delay the eventual temperature increase. But the rising ocean level so caused would present an even more serious problem to the future of humanity.

The author acknowledges interesting discussions with A. Berger, M. I. Budyko, Walter Elsaesser, H. Flohn, M. Ghil, J. T. Hollin, W. W. Kellogg, G. Kukla, H. L. Huo, Syukuro Manabe, Philip Morrison, Ralphy Rotty, R. H. Thomas, and Alvin Weinberg.

References

[1] Hays, J. D., Imbrie, J., and Shackleton, N. J.: 1976, *Science* 194, 1121.
[2] Imbrie, J. and Imbrie, J. Z.: 1980, *Science* 207, 943.
[3] North, G. R. and Coakley, J. C.: 1979, *J. Atmos. Sci.* 36, 1189.
[4] Pollard, D.: 1978, *Nature* 272, 233.
[5] Pollard, D., Ingersoll, A. P., and Lockwood, J. G.: 1980, *Tellus* 32, 301.
[6] Schneider, S. H. and Thompson, S. L.: 1979, *Quat. Res.* 12, 188.
[7] Suarez, M. J. and Held, I. M.: 1979, *J. Geophys. Res.* 84, 4825.
[8] Berger, A. L.: 1977, *Nature* 269, 44.
[9] Sellers, W. D.: 1969, *J. Appl. Meteor.* 8, 392.
[10] Berger, A. L.: 1977, *Palaeogeog. Palaeoclim. Palaeoecol.* 21, 227.
[11] CLIMAP Project Group: 1976, *Science* 191, 1131.
[12] Weertman, J.: 1976, *Nature* 261, 17.
[13] Birchfield, G. E.: 1977, *J. Geophys. Res.* 82, 4909.
[14] Birchfield, G. E. and Weertman, J.: 1978, *J. Geophys. Res.* 83, 4123.
[15] Källén, E., Crafoord, C., and Ghil, M.: 1979, *J. Atmos. Sci.* 36, 2292.
[16] Sergin, V. Ya.: 1979, *J. Geophys. Res.* 84, 3191.

[17]   Schneider, S. H. and Dickinson, R. E.: 1974, *Rev. Geophys. Space Phys.* 12, 447.
[18]   Budyko, M. I.: 1969, *Tellus* 21, 611.
[19]   Held, I. M. and Suarez, M. J.: 1974, *Tellus* 26, 613.
[20]   Ghil, M.: 1976, *J. Atmos. Sci.* 33, 3.
[21]   Fredericksen, J. S.: 1976, *J. Atmos. Sci.* 33, 2267.
[22]   Su, C. H. and Hsieh, D. Y.: 1976, *J. Atmos. Sci.* 33, 2273.
[23]   Saltzman, B.: 1977, *Tellus* 29, 205.
[24]   Kraus, E. B.: 1980, private communication.
[25]   Weertman, J.: 1964, *J. Glaciol.* 6, 145.
[26]   Cess, R. D.: 1976, *J. Atmos. Sci.* 33, 1831.
[27]   Goldthwait, R. P.: 1958, *Ohio J. Sci.* 58, 209.

(Received September 24, 1980; in revised form May 18, 1981).

Origin of Ice Ages: Initial Condition Forcing and Dynamics

Peter Fong

Physics Department, Emory University

Atlanta, Georgia 30322, USA

Abstract

Ice ages happened only in Pleistocene. The origin of ice ages is thus hinged on the initial condition of the Pleistocene environment, which is based on a polar centered continent with a permanent, constant, equilibrium (with the ocean) icesheet on it. The Milankovitch eccentricity forcing will perturb the equilibrium condition and thus cause glacier advance and retreat of the icesheet, which would be small by the current climate theory based on stable equilibrium. We prove that the equilibrium is not stable, but neutral, due to ice-water phase equilibrium on iceshelves. Hence the glacier advance will be large, like a cylinder rolling away to infinity by a weak forcing. Thus the Pleistocene icesheet will be the womb to spawn the ice age and only in Pleistocene. This being the necessary and sufficient condition of an ice age, no other causes are relevant.

1. Introduction

There are three major problems concerning the ice ages. First is a large amount of ice accumulation on the continents with a 10^5 year cycle. Second is a reasonable amount of cooling of the earth. Third, these drastic changes happened only in Pleistocene (and possibly 300 and 600 million years ago) in 1% of the geological time.

In the past century several dozens of theories have been advanced on the origin of ice ages considering problem 1 or 2 or both, but practically none on problem 3. There were impressive results obtained[1] but their very successes were their undoing because they would also predict ice ages before Pleistocene, which are unwanted and doom the theories.

Whatever forcing involved in the ice age thus cannot be a strong one, which would force ice ages to appear before Pleistocene. It follows necessarily that there must be an "amplification" mechanism to enhance the weak forcing by the Pleistocene environment which will be studied.

A favorite cause of ice age considered before is the Milankovitch forcing due to the change of the eccentricity of the earth orbit, which has the right periodicity of 10^5 years. The variation of insolation due to the eccentricity changes ranges from -0.17% to +0.014% according to Berger.[2] The magnitude of this energy change has been shown by Sellers[3] and others to be too weak to generate the temperature change of the ice ages (problem 2) by a factor of 10. Hence this forcing has been rejected.

While this forcing is 10 times too weak for Problem 2, it is 10 times more than necessary for Problem 1. The insolation deficit due to the eccentricity change of the earth orbit is more than enough to remove the heat of fusion and create 3.5×10^{22} gm of ice of an ice age. However, the ice formed in winter will melt in the summer and will not accumulate to form an ice age. (This correctly explains no ice ages before Pleistocene.) On the other hand during Pleistocene there is a permanent icesheet in Antarctica, which does not melt in summer. The ice formed can accumulate on it to form an ice age. Once Problem 1 on the origin of ice is solved, Problem 2 on temperature can be solved as an after effect—the earth cools down due to the extra albedo of the new ice. This solution of Problem 2 being correct and unavoidable, Sellers' theory is like a second bride in a wedding that is an unwanted intruder. Indeed it is rightly a useless spoiler. Melankovitch forcing is thus likely to be the right forcing.

2. The Qualitative Theory

We develop a qualitative theory of ice ages starting with the origin of the initial condition of the Pleistocene icesheet. About 15 million years ago the Antarctica continent has drifted to the present polar position. Ice accumulated on it to form a permanent icesheet which expanded to cover the entire continent and reached the ocean, forming iceshelves. Equilibrium with the ocean was established through the iceshelves, which stabilized the icesheet size(with minor seasonal variations of the iceshelves). This seems to be the initial condition from which an ice age started.

Now the uninvited but ever-present Milankovitch forcing will naturally perturb the equilibrium condition of the icesheet, generating glacier advances and retreats, just like seasonal perturbations generating iceshelves advance and retreat of hundreds of meter in a yearly cycle.

However, according to the current climate theory, such as Sellers,[3] which is based on the climate system being in stable equilibrium, the effect would be small and negligible. The Milankovitch forcing, after generating a small effect, will reach a new equilibrium and cause no further change. That the ice age grows out of the Pleistocene icesheet is an indisputable fact. The only unexplained point is the large extent of the glacier advance.

The large amplification mechanism needed cannot happen in a *stable* equilibrium (and in most conventional meteorology) where the response to a small perturbation is also small. But it can occur in unstable equilibrium of touch-and-go situations like the tumbling of a precariously balanced cone on its tip. On the other hand such changes take place in a short time with catastrophic results, which is like a tornado, not like a glacier advance. What we need here is the weakest kind of unstable equilibrium, that is, the *neutral* equilibrium, which involves a small change that lasts for a long time, during which a large effect may be accumulated.

An equilibrium ellipsoid under a light push will tilt a little, reaching a new stable equilibrium and stop there. The new equilibrium state, different from the original one, generates a restoring force to balance off the external forcing and stop further changes. A sphere, which is in neutral equilibrium, under a light push will not stop in a new stable equilibrium but will roll into a new neutral equilibrium, and keeps on rolling into new neutral eqilibria *ad infinitum*—the sphere will run away to infinity. The reason is that the new equilibrium state is the same as the old (thus neutral) and does not generate a new restoring force to balance off the external forcing, which thus continues to push the system forward, and forward,..., *ad infinitum*. This way we have the amplification effect of a neutral equilibrium we are looking for. If we can prove the Pleistocene icesheet is in neutral equilibrium then we have a good explanation of the origin of ice ages. (Before Pleistocene there was no icesheet and the system was in stable equilibrium; and we have no ice ages.)

Now we proceed to prove that the Pleistocene equilibrium icesheet is indeed in neutral equilibrium, not in stable equilibrium, which has never been proved, only carelessly assumed. Actually this is merely a natural and unavoidable conclusion of thermodynamics. The Antarctic icesheet is in equilibrium with the ocean through the iceshelves. The physical basis of iceshelves-ocean equilibrium is the phase equilibrium of ice and water, which, by the laws of thermodynamics, is a neutral equilibrium (the total entropy change in a reversible phase transition is zero $dS=0$ signifying neutral equilibrium, not greater than zero $dS>0$ as in stable equilibrium).

We can now explain the large amplification effect of the Pleistocene icesheet in generating an ice age. In Sellers argument in the absence of an icesheet, the Milankovitch cooling forcing will generate a new stable equilibrium of the climate system with a lowering of temperature of a few tenths of 1°C. The reduction of infrared heat loss of the earth due to the cooling then balances off the Milankovitch cooling forcing and the earth will cool no further. (This explains no ice ages before Pleistocene.)

With the icesheet in Pleistocene, the forcing will, and will only, cause the freezing of some ice to be added to the equilibrium icesheet but not change the temperature (always at 0°C) according to the law of phase equilibrium. The icesheet system then moves into a new neutral equilibrium state, which is exactly the same as the old state, and will cause changes as the old state, and this will be repeated *ad infinitum* leading to the accumulation of a large amount of ice on the Pleistocene icesheet resulting in a large glacier advance. We have already mentioned that the Milankovitch forcing is large enough to generate all the ice of an ice age. This explains *fully* the Problem 1. Here the Pleistocene icesheet acts as the womb to spawn the ice age; the only thing unusual is the large size of the fetus (150% of the size of the mother) which is due to neutral equilibrium.

Now the after effect of the ice increase. The large amount of ice generated will in turn lower the temperature of the earth. This will explain Problem 2, which is a separate issue, involving separate laws of nature and a larger physical system (the surface of the earth) with a new independent variable T, the temperature of the earth (essentially the ocean).

The essential physics in the larger earth surface system is this: As ice is added to the growing Pleistocene icesheet, it generates more reflecting surface and dissipates more heat to outer space (albedo positive feedback D_p). As a result the earth (essentially the ocean) cools down (infrared negative feedback D_n) and finally reaches equilibrium ($D_p=D_n$).

The earlier smaller system is in neutral equilibrium, leading to a continuous sequence of neutral equilibrium points. This sequence of points is inherited in the larger system, which is thus also in neutral equilibrium.

Now the last problem of the termination of the glacier advance. The rolling of a sphere in neutral equilibrium will approach infinity. Would the glacier advance approach infinity (the equator)? Having solved the difficult problem of getting the glacier advance started, and made sure it will keep on advancing, we now face an equally difficult problem to stop the advance because we know that glacier never advanced to the equator, which would be a disaster. There are theories succeeded in getting glacier advance started but will drive it all the way to the equator. The stop of a rolling sphere requires an external factor; so is the glacier advance.

It can be shown that a two-dimensional icesheet will indeed advance to the equator but a three dimensional icesheet will not. The physical reason is that in a three dimensional icesheet in the late stage of its growth, new snow will largely pile up on old snow instead of on barren ground to generate new reflecting surface. The albedo positive feedback will be reduced, which slows down the glacier advance and eventually stops it. This way every aspect of the ice age problem can be accounted for qualitatively. The quantitative details will be given in the nexe section.

3. The Quantitative (Dynamic) Theory

The above qualitative theory is complete by itself. It is desirable to put it in mathematical form so that the results can be verified quantitatively. In particular, to solve the problem 2 and the problem of the termination of glacier advance at about 40°N latitude, not at the Equator.

The ice age dynamics is a quasi-static process of slow changes along a sequence of neutral equilibrium points of the earth surface system driven by a small external forcing which breaks the neutral equilibrium of the system. The dynamics thus contains the Equations (1 and 2) below of the mass and energy conservation of the glacier advance processes as well as the neutral equilibrium conditions of the iceshelves Eq.(1) and the earth surface Eq.(3). From them we try to establish a dynamic theory of the ice age glacier advance specified by the two variables $W(t)$, the total icesheet mass (Antarctica plus Greenland), and $T(t)$, the ocean surface temperature. The $W(t)$-$T(t)$ system contains less information than the general earth surface system. Together with the quasi-static condition, the above mentioned equations may be sufficient to establish the deterministic dynamic equations to solve the ice age problem. The system is perturbed in general by a heat deficit $D(t)$ cal/year which can be split into two parts: the latent heat $D_L(t)$ and the sensible heat $D_s(t)$.

1. Mass conservation equation of the iceshelve system

$$\frac{dW(t)}{dt} = \frac{D_L(t)}{L_f} \frac{gm}{yr} \qquad (1)$$

where $L_f = 80$ cal. This is also the neutral equilibrium condition of the iceshelves system on phase transition because only at neutral equilibrium water converts to ice by the above equation.

2. Energy conservation of the ocean in the earth surface system

$$\frac{dT(t)}{dt} = -\frac{D_s(t)}{C_m} \, ^\circ C \, yr^{-1} \qquad (2)$$

where C_m is the heat capacity of the ocean mixed layer, assumed to be 200 meters thick.

3. Neutral equilibrium condition of the earth surface system

$$D_p = D_n \qquad (3)$$

Neutral equilibrium is broken by the external forcing D_e in glacier advance

$$D_s + D_L \equiv D = D_e + D_p + D_n \qquad (4)$$

4. An equation implying Eq.(3) as its substitute for convenience of calculation. For one period of glacier advance of an ice age D_p is generated by an amount of ice ΔW corresponding to D_L equal to 3.1×10^{24} cal by Eq. (1), and D_n is generated by an amount of ΔT corresponding to a D_s equal to 1.7×10^{23} cal by Eq. (2). Equation (3) implies that D_s and D_L are related by the ratio of $1.7 \times 10^{23} \div 3.1 \times 10^{24} = 0.053$

$$R \equiv D_s/D_L = 0.053 \qquad (5)$$

From the sum (4) and the ratio (5) of the two quantities D_s and D_L, we can calculate the two and use them in Eqs. (1,2) to calculate $W(t)$ and $T(t)$.

5. The albedo and infrared feedbacks are calculated from standard meteorology

$$D_p(t) = \int_{A_0}^{A(t)} (\text{solar constant})\, \Delta(\text{albedo}) \, \frac{\cos \ell}{\pi} \, dA(t) \qquad (6)$$

where ℓ is the latitude and $A(t)$ is the ground area of the three dimensional icesheets projected on the earth surface,

$$D_n(t) = \int -0.7\, I\, \frac{1}{\beta} \frac{dT}{dt}\, dt \qquad (7)$$

where I is the annual isolation, 0.7 is the fraction of I in the earth heat income compensated for by the infrared radiation. ß is the infrared sensitivity parameter equal to 145°C.

6. The geometry of the three dimensional icesheets of the two hemispheres is assumed to be two identical circular cones of a radius-to-height ratio of 1000. $W(t)$ and the radius $r(t)$ of the area $A(t)$ are related by

$$W(t) = 0.917 \times 2 \times \tfrac{1}{3}\, \pi r^2\, \frac{r}{1000} \qquad (8)$$

From $r(t)$ we obtain the area $A = 2\pi r^2$ for Eq. (6).

Equations (1-8) determine the independent variables $W(t)$ and $T(t)$ completely and uniquely. Thus they *are* the dynamic equations of the $W(t)$-$T(t)$ system which we desire. The difficult dynamic problem is solved through the back door, thanks to all the favorable conditions to finesse it.

Once the initial conditions $W(0)$ and $T(0)$ of the Pleistocene icesheet are specified and the external forcing $D_e(t)$ is identified with the Milankovitch, the equations can be integrated by computer and $W(t)$ and $T(t)$ will explain the glacier advance to solve all Problems 1, 2 and 3.

Typical solutions based on a constant Milankovitch forcing of 0.13% of the insolation leads to glacier advance and earth cooling histories plotted in Fig.1 in agreement with the observed results of ice volume and temperature reduction at maximum glaciation. Noteworthy is the result that the glacier advance stops at the middle latitudes around 40°N, not at the equator, the glacier termination being a sensitive test of the theory.

Solutions based on a sinusoidal Milankovitch forcing of the same strength show nearly sinusoidal glacial advance and retreat as expected. Noteworthy are the results that there is no phase shift of the motion cycle from forcing cycle and there is a hysteresis effect of 22.4% ice retentivity.

4. Meaning of the Mathematical Results

The computer printouts of the integration of the differential equations (1-8) include the results of the intermediate variables $D_p(t)$ and $D_n(t)$, the positive and negative feedbacks. They are plotted in Fig. 2. It is interesting to note that the two are nearly equal at all times. This mathematically proves that neutral equilibrium is nearly established all the time as it should because of the phase equilibrium of ice and water at the iceshelves. This point is not introduced *a priori* in the beginning of the mathematical theory but derived *a posteriore* at the end.

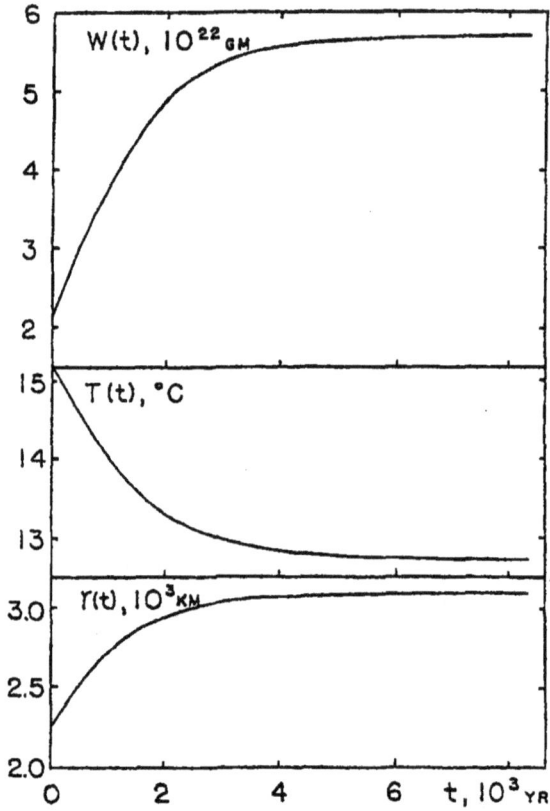

Fig. 1. The evolution of the ice age glacial system predicted by he dynamical theory. $W(t)$ is the total mass of the icecaps. $T(t)$ is the average ocean surface temperature. $r(t)$ is the average radius of the icecaps. All are functions of time t.

A closer inspection of Fig. 2 reveals that D_p and D_n are not exactly the same with D_p slightly greater than D_n in the beginning and then smaller later all the way to the end. The difference D_p-D_n in addition to D_e is what drives the advance of the glacier from one equilibrium point of the icesheet to another and yet another,...., in a sequence of changes forming a quasi-static process. The reverse of the sign of D_p-D_n later means the slowing down of the advance which is eventually brought to a stop in equilibrium when $D_n = D_p + D_e$ as shown in Fig. 2. At this point the external forcing is balance off by the feedbacks and the advance stops at maximum glaciation.

In the Sellers theory based on stable equilibrium, once D_e is switched on at the beginning of the curve in Fig. 2, the stable equilibrium of the system *generates a negative feedback* canceling out D_e immediately and the curve stopped dead there with no continued glacier advance. In the present theory the external forcing D_e *generates some ice* according to the law of phase equilibrium. The ice then generates positive and negative feedbacks that are balanced off in the free-wheeling state of neutral equilibrium and therefore the external forcing can continue to drive the glacier advance without hindrance of restoring force in a massive ice advance. The change of sign of D_p-D_n as time goes on and as the icesheet grows is determined by the geometric effect of the three dimensional icesheet as explained earlier. The effect also brings the glacier advance to a stop at middle latitudes.

We have developed a general method to treat cryosphere interaction in a more general climate system of atmosphere-hydrosphere-cryosphere characterized by neutral equilibrium beyond the conventional climate system of atmosphere-hydrosphere characterized by stable equilibrium.

As an application the theory is just what needed for the greenhouse warming problem which includes the cryosphere. The global warming

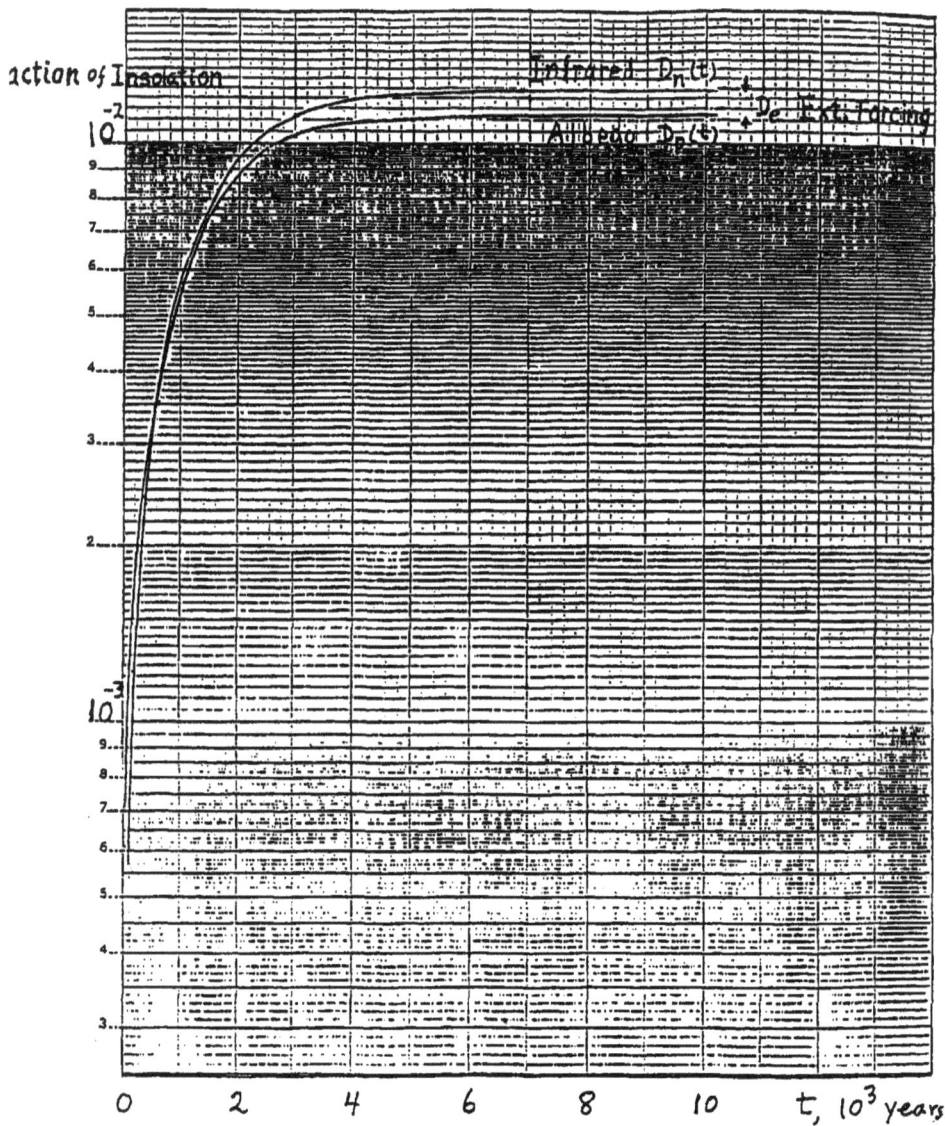

Fig. 2. Positive and negative feedback as functions of time.

process is the same as ice age glacier retreat. Thus Eq. (5) deems that the heat of earth warming is 5% of that of ice melting. The fact today is that the iceheet is not melting and the heat of melting is zero. 5% of zero is also zero. Thus the greenhouse warming is zero now which is verified by facts.

The miracle of this work that has eluded all past geniuses is hidden in the *discontinuity of the equation of state* displayed in phase transition. This breakthrough can be extended to another phase transition to unlock the secret of the future of greenhouse warming. Result: No greenhouse warming until the sky is entirely covered with clouds 3000 years later.

5. Afterthoughts

The set of equations (1-8) obtained here turn out to be the same as the equations (1-8) used in a paper of mine[4] published in 1982 except minor simplifications. (Many of the detailed discussions of the equations in the 1982 paper are thus not repeated here.) That paper on the same subject of the origin of ice ages was published before the crucial idea of neutral equilibrium was introduced. That such a feat was possible at all must be considered as a miracle. It reminds one of the comment of Nobel Laureate P. A. M. Dirac on the equation that bears his name (Dirac relativistic wave equation) that the equation is more intelligent than its author.

In the present case the 1982 paper generated such awe and disbelief that the Editor felt necessary to add a Note of disclaimer at the bottom of the title page. The paper has been treated as an untouchable if not a pariah. It was the attempt to find a more congenial approach that led to a series of developments[5] culminated in the discovery of the role of neutral equilibrium, highlighted by Fig. 2, which was not plotted until 1990.

In retrospect it seems that the earlier paper, though without the correct guiding principle, has hit the anatomy of the ice age right and was able to produce the right anatomy of the body in Figure 1 even without a soul. Besides, there were technical accomplishments that no one could deny (the introduction of latent heat into the climate theory, of the deep ocean to replace the shallow ocean, and of the three dimensional icesheet). With the discovery of the results of Figure 2 suggesting neutral equilibrium, the corpse was seized by a voodoo god and turned into a zombie, which was still to be feared and desisted. Finally a soul was found in the idea of neutral equilibrium and the theory was worked out, as presented in this paper, in the pedestrian and yet canonical Newtonian formulation with homespun and yet unchallengeable initial condition, forcing and dynamics without any sense of mystic Revelation. It is now a kind and amicable being and the 16 years of wandering in the desert has ended.

In contrast the ice age theories in the past century have the initial condition, forcing and dynamics presented as a bark up the wrong tree. There was nothing even tangentially relevant. The search of the origin of ice ages has extended to such esoteric heavenly oddities as the earth orbit plane oscillations and the interaction of the moon orbit with the tide.

References

1. See, for example, Lev Tarasov and W. R. Peltier, EOS, Trans. Amer. Geophys. Union, **78**, #46 Supplement, F22 (1997), and JGR-Atmosphere, September 1997.

2. A. L. Berger, Nature **269**, 44 (1977), and Palaeogeog. Palaeoclim. Palaeoeclo. **21**, 227 (1977).

3. W. D. Sellers, J. Appl. Meteor. **8**, 392 (1969).

4. Peter Fong, Climatic Change, **4**, 199 (1982).

5. Peter Fong, EOS, Trans. Amer. Geophys. Union, **64**, 202 (1983), **64**, 671 (1983), **68**, 1234 (1987), **71**, 1262 (1990), **75**, #44 Supplement, 213 (1994), **76**, #17 Supplement, S171 (1995), **76**, #46 Supplement, F305 (1995), **78**, #46 Supplement, F24 (1997).

Unraveling a Century's Mystery of the Ice Ages

Peter Fong

Physics Department, Emory University

Atlanta, Georgia 30322

Abstract. *The origin of ice ages includes: (1). The polar position of Antarctica due to continental drift. (2). Its permanent icesheet has iceshelves formed neutral equilibrium with the ocean. (3). This icesheet grew by the Milankivitch eccentricity forcing and acted as the seed and womb to spawn new icesheet. (4) The icesheet grew continuously without stop because in neutral equilibrium there is no restoring force to stop it, leading to a free-wheeling action creating a very large icesheet. (5). The large ice generated will lower the temperature of the earth, overcoming the difficulty for the eccentricity forcing to do it directly. Any 1 of the 5 or more could be easily overlooked leading to a mystery lasting for a century.*

After a century, the search for the origin of ice ages has turned to even more esoteric happening in the universe such as the oscillation of the earth orbit plane and the interaction of the moon orbit with the tide. The deepening of the mystery is partly due to the wrong directions of pursuing the initial condition and forcing. The first is that the Milankovitch eccentricity forcing was dismissed for being too small. But it may be the only right one for being able to account for the 2.8×10^{24} calories of heat of fusion needed in an ice age; the ice will then lower the temperature of the earth, creating an ice age. But then why did not this forcing generate ice ages before Pleistocene? The second error is to miss a necessary initial condition of the existence of a permanent icesheet (that of the Pleistocene), which has important bearings on this and another aspect of the study of ice ages. The traditional approach is to find a mechanism to lower the temperature of the earth (the eccentricity forcing is too weak for that), which hopefully will then generate the ice of the ice age. However, whatever ice generated will be lost in the summer heat unless there exists a permanent icedsheet at 0°C year around to act as a womb to protect the ice and let it grow. With the said initial condition the icesheet can grow to generate an ice age. Without that initial condition, there can be no ice ages. That is why there was no ice ages before Pleistocene. Since the large amount of ice generated will cool down the earth by the ice-box effect, there is no need of a Herculean effort to find a mechanism to cool down the earth in the first place as in the traditional approach.

The traditional approach suffers from the intriguing mystery that the ice age is difficult to start, but once started it is difficult to stop—it will go all the way before Pleistocene and go all the way to the equator of the earth. Both did no happen. To start and stop at will is an added problem.

In an ice age the Milankovitch eccentricity forcing provides an energy deficit of about 9×10^{24} cal, enough for the creation of 3.5×10^{22} gm of ice. No other theories of ice ages have accounted for this basic fact of ice calorie for calorie to reach the first base, let alone for the home run.

To reach the second base we establish the necessary initial condition of an equilibrium icesheet which is provided for us by the Pleistocene. Its ultimate origin is the continental drift which brought Antarctica continent to the polar position 15 million years ago. The cooler polar climate created and accumulated ice to form an ice mass in year-around presence. Glacier flow then established an *equilibrium icesheet* with iceshelves in equilibrium with the ocean.

Then the Milankovitch eccentricity forcing (in the cooling half cycle) would act on the equilibrium icesheet The equilibrium point is the iceshelf-ocean inter-face and the nature of equilibrium is phase equilibrium of ice and water, which, by the laws of thermodynamics, is a *neutral equilibrium* (dS=0, instead of dS>0), not different from the case of ice cubes in a cold drink. During the cooling half cycle a negative energy perturbation is applied and the neutral equilibrium system must respond by converting some water into ice without changing the temperature, which is the phase transition temperature of 0°C, just as what would happen to ice cubes in a cold drink placed in a freezer. The resulting new system is in the same neutral equilibrium state of the old (by definition of neutral equilibrium) and the same change will be repeated, as long as the forcing supplies the heat deficit, generating a continuous series of neutral equilibrium states, forming a free-wheeling action like a sphere on a plane rolling to infinity. It will stop after the cooling half cycle is over, creating an ice mass of 3.5×10^{22} gm of the ice age as we have calculated earlier. (In *stable equilibrium*, a small push on an *ellipsoid* will instead generate a small tilt together with a restoring force that will stop further movement.) The free-wheeling action generates a large amount of ice from a small forcing because it is over a long time of the ice age, in which the weak forcing effect can add up to a large amount of ice. This explains how the weak Milankivitch forcing can generate a large glacier advance under the *neutral* equilibrium condition. Incidentally conventional climate theories are based on *stable* equilibrium and missed the essence of the ice ages.

In the meantime the increased ice generates increased albedo (positive feedback) and cools down the earth. The lowered temperature of the earth leads to reduced infrared radiation (negative feedback) of the earth which balances off the positive feedback and restores the balance of the energy budget of the earth. Thus each equilibrium point at the iceshelf-ocean boundary leads to an equilibrium point of the *earth system* as a whole characterized by the equality of the positive and negative feedbacks. Therefore the series of free-wheeling points at the iceshelf-ocean boundary generates a series of points of the earth system as a whole, which are also free-wheeling (the equality of positive and negative feedbacks is another definition of neutral equilibrium). Thus the earth climate system is also in *neutral equilibrium* and the small Milankovitch forcing can induce large change of the climate, this time it includes the change of the temperature as well. Previously it was thought Milankovitch forcing was too small to give rise to the large change of temperature in an ice age. This is true in an ice free earth in stable equilibrium (as the earth before Pleistocene). The existence of the *equilibrium icesheet* changes the situation completely. From the neutral equilibrium condition of the equality of the albedo and infrared feedbacks we can calculate the lowering of temperature of the earth from the amount of ice accumulated. The result agrees with what has been actually observed.

The contrast of neutral with stable equilibrium can be illustrated by the climate system before Pleistocene which, without an icesheet, is in *stable equilibrium*. The ever-present Milankovitch forcing will now act on the stable equilibrium system. If it creates some ice it will melt in the summer, there being no permanent icesheet present in the summer, providing an environment of 0°C for the Milankovitch ice to survive through the summer and to accumulate to form a large icesheet. If it cools down the earth, the action in a stable equilibrium will generate a restoring force canceling out the forcing and stop further change of the temperature.

The net results are an insignificantly small change of the ice and temperature. Thus before Pleistocene there can be no ice ages for the very same Milankovitch forcing. In contrast in neutral equilibrium there is no restoring force to stop further change, and the applied perturbation can act continuously to generate a large change.

Thus the common errors in previous theories on the initial condition (omitting the Pleistocene icesheet) and forcing (neglecting Milankovitch forcing) can be corrected on the basis of new knowledge developed by the study of neutral equilibrium. And the ice age can be explained.

If the neutral equilibrium free-wheeling action were to continue without end, the glacier advance would progress all the way to the equator, which would wipe out all lives on the earth. This should happen but did not, a happy mystery, which has not been explained. Now we have the explanation in the three-dimensionality of the icesheet (see later). A two-dimensional icesheet would indeed progress to the equator.

Once stopped in the middle latitudes, the icesheet waited for the warming half cycle of the eccentricity forcing to change glacier advance to retreat and to return to the starting point of the cycle in an interglacial. The residual icesheet of the interglacial now provides the necessary initial condition for the next ice age and the cycle repeats *ad infinitum*. If the glacier were to advance to the equator, it would get stuck there forever, and there would be no repeated ice ages.

The ice age cycles would terminate should the Antarctica continent move out of the polar position by future continental drift in hundreds of million years. The grand cycle of in-and-out of the polar position of a certain continent might have happened before in 300 and 600 million years ago resulting in batches of ice ages, the existence of which seems indicated by some geological evidence.

In retrospect it seems clear whatever forcing of the ice age must be a weak one so that it can be manifested in nothing before Pleistocene but it can be "amplified" to become very large by a special mechanism in Pleistocene. In stable equilibrium the cause and effect are of the same order of magnitude and there can be no amplification. In unstable equilibrium such as the toppling of an upright pole, a weak cause (trigger) can indeed generate a large effect, but in a brief catastrophe with a big bang. The ice age phenomena proceed slowly and peacefully for a long time without a big bang, which are typical results of neutral equilibrium. Without going into details it is clear neutral equilibrium is the essence of the ice ages, which was ignored in a century.

The explanation of ice ages in this paper can be expressed in a mathematical form as a dynamic theory. Indeed a mathematical dynamic theory of glacial advance and retreat has been formulated and published[1] in 1982 based on Milankovitch eccentricity forcing with results in agreement with the observed ice ages. But that was before the discovery of the neutral equilibrium theory. It was amazing that without the soul of the neutral equilibrium theory a viable theory could have been developed at all, like a cart moving without a horse. What has been developed in that paper was a respectable autopsy report of the ice age which brought out the complete dead anatomy without the controlling soul. A skillful forensic lawyer could have spun out a lively tale based on an autopsy report; but it will not be taken as evidence in the court. However, if crucial evidence is found to support it, the lively tale may be confirmed as the truth. Before that, a soulless body can hardly be regarded as humankind.

To establish the connection of that theory with the neutral equilibrium theory I have retrieved the unpublished data of the positive and negative feedbacks of the earth climate system of the 1982 paper from the computer printout and plotted them here in Figure 1. It shows indeed the two feedbacks are nearly equal as a characteristic feature of neutral equilibrium. [The figure also shows that it gradually changes into stable equilibrium (negative infrared > positive albedo) and stops the glacial advance in middle latitudes (not in the equator) due to the three dimensionality of the icesheet.] Thus neutral equilibrium was actually *derived* from the 1982 theory, which did not embody it as the missing soul and was severely criticized for the transgression. Now it happens in logic often that in many cases the conclusion can be used to derive the premises of the theory. Thus the truth of neutral equilibrium is actually contained in the 1982 theory—the soul was already there, only hidden behind a veil. The theory is thus redeemed in hindsight and can be enjoyed as a forensic stunt. This completes the home run.

References

[1] Peter Fong, Climatic Change, 4, 199 (1982).

Figure Caption

Figure 1. Positive and negative feedbacks as functions of time of an earth climate system representing a glacier advance. The equality of the two in the early growth years indicates the establishment of a neutral equilibrium in a free-wheeling action in which a small forcing can generate an extremely large change of the ice volume and temperature. Toward the end the equilibrium changes back to stable due to the three-dimensional effect of the icesheet and stops the glacier advance way before reaching the Equator.

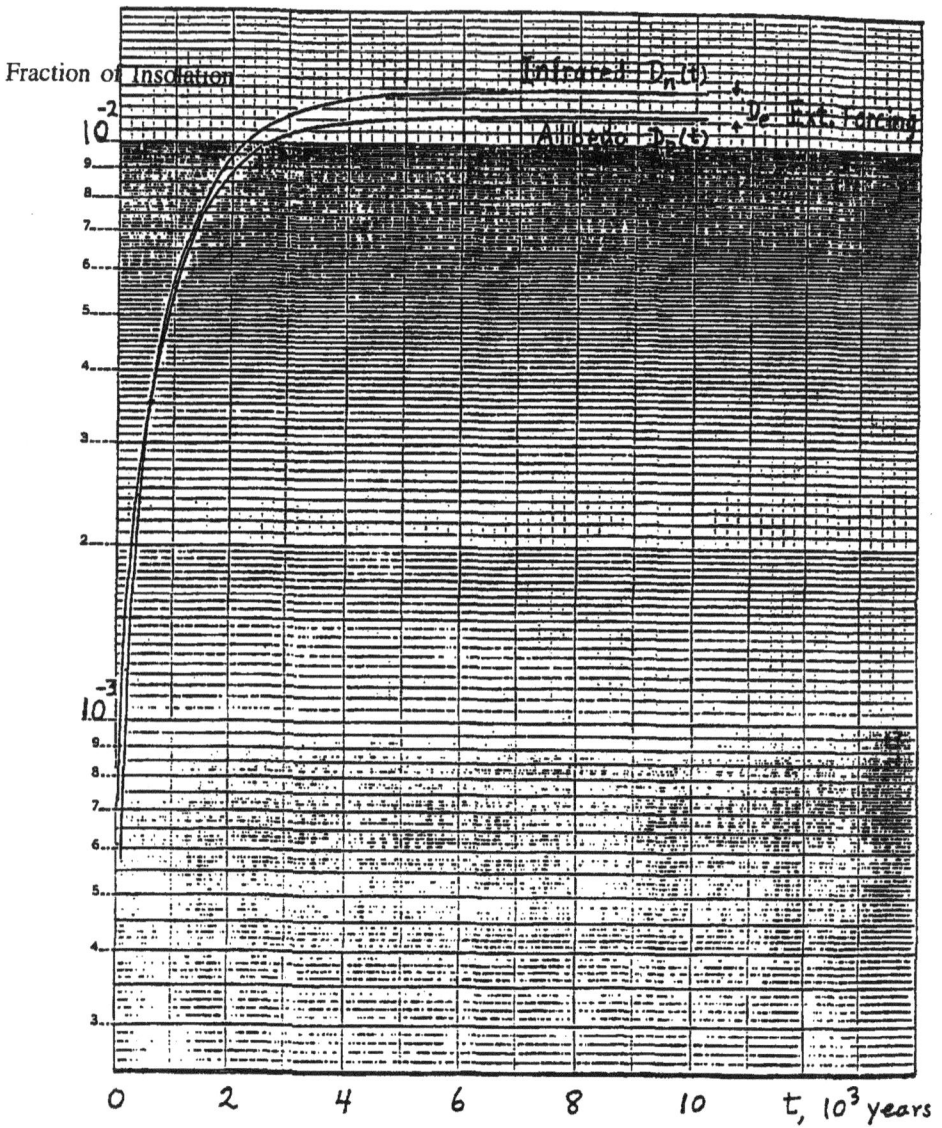

Fraction of Insolation

10^{-2}

9
8
7
6
5
4

3

2

10^{-3}

9
8
7
6
5

4

3

Infrared $D_n(t)$

Albedo $D_a(t)$

$1/e$ with Forcing

0   2   4   6   8   10 $t, 10^3$ years

Figure1

The Fourth Phase of Water and the Theories of Greenhouse Effect and Ice Age Glacial Cycles

Peter Fong

Physics Department, Emory University

Atlanta, Georgia 30322, USA

Besides the three phases of water, solid, liquid and gas, this paper introduces the aerosol phase as in a cloud as the fourth phase. The transition of cloud water droplets to saturated vapor can be shown to take place at a constant temperature equal to the dew point with a latent heat around 590 cal, thus a bona fide phase transition. Hence cloud is a thermostat just as polar ice and ocean water forming a thermostat. These two thermostats tend to maintain a constant temperature of the earth in the interglacial period, which we are now in. Under this premise the global heat balance (including albedo and infrared feedbacks) leads to 2 consequence: (1) In the short run there is no global warming; the anthropogenic greenhouse heat generates increased cloudiness and precipitation but the increased albedo dissipates exactly the greenhouse heat thus generates zero warming. (2) In the long run the Milankovitch forcing perturbs the neutral equilibrium system of the 2 thermostats and generates glacier advance and lowering of earth temperature in the ice age. These two predictions are verified by mathematical calculations in other papers. Actually greenhouse warming is a special case of the theory of ice ages in which the cloud thermostat generates

the condition of no warming and thus no infrared feedback, and the global equilibrium condition (ultimately traceable to the second law of thermodynamics) determines that the albedo feedback removes the greenhouse heat completely, leaving no greenhouse warming but increased cloudiness and precipitation as the only observable and meaningful greenhouse effects.

The aerosol phase is defined as water in the droplet form suspended in air so small that they will not precipitate but will be in equilibrium with the saturated vapor to form clouds. A collection of droplets represent a molecular aggregate structure distinctive and discontinuous from those of the three other phases and thus qualify as a separate phase in the most general definition of the phase.

However, the aerosol phase so defined is stable only when in equilibrium with the saturated vapor, another phase, in the cloud. The two phases mix homogeneously to form a cloud and are not physically separable. Therefore the aerosol does not qualify as a phase as used in the Gibbs phase rule. However, phase rule represents only one aspect of the phase transition phenomena and the conventional phase definition may be considered a special kind dedicated to the Gibbs phase rule. It may be called the Gibbs phase. We are at liberty to define a more general phase to deal with phase phenomena beyond Gibbs phase rule.

Ordinarily we consider one state of matter (gas, liquid or solid) as one phase. But the solid state of water ice has 4 crystal forms, thus 4 solid phases. The taboo of single phase is broken which helps us here for the cloud phases, the number of which is actually infinite because the cloud can have infinitely many compositions of water droplets. Moreover, these phases are continuous, not discrete (as ice phases), which is beyond Gibbs' phase rule. Furthermore, the phases are metastable. There is not a laboratory that has a case showing a permanent phase of a cloud forever. Once the taboo is broken, there is no end. We can define general phases in any way we wish to fit our needs.

Phase transition. The aerosol phase may change to the vapor phase when a cloud evaporates and disappears as heat is added. It may also change to the ordinary water phase when a cloud precipitates into rain drops, which are no longer *suspended in air*. In both cases, discontinuous changes of the molecular aggregate structure take place, fulfilling the definition of phase transition.

Latent heats. The latent heat in the first case is the heat of vaporization of water at the temperature of the cloud (the dew point), which is heat of vaporization at the dew point temperature of the cloud, about 590 cal/gm depending on the dew point. The phase transition is thus of the first order. In the second case the latent heat is zero and the process is phase transition of the second order.

To qualify as phase transition as conventionally defined, it requires to show that the temperature remains constant as the transition proceeds, so that all heat perturbations during the transition go to latent heat and none to the sensible heat. Ordinarily this is demonstrated by experiment and phase transition is an empirical phenomenon. Here we choose to prove it theoretically. Also there is no suitable cloud specimen to experiment with.

Theorem. The phase transition from aerosol to saturated vapor takes place at a constant temperature which is the dew point of the cloud (about 20°C depending on the cloud). (This temperature may be called the "boiling point" of the aerosol phase which is distinctively lower than that of the ordinary water phase 100°C. This distinction turns out to be crucial.)

Proof. Let a beam of radiation shine on a cloud. The energy delivered to the water droplets is divided into two parts: one is delivered to the volume of the droplet as sensible heat to raise its temperature and the other to the surface as latent heat to evaporate the droplet. The volume energy is proportional to the volume and thus to r^3 where r is the radius. The surface energy is proportional to the surface and thus to r^2. The ratio of volume energy to surface energy is proportional to r, which approaches zero when the droplet radius approaches zero in the aerosol phase. Thus in the aerosol phase all heat perturbations are taken up by latent heat. There is no sensible heat to increase the temperature. Therefore, the temperature is a constant, which is the dew point of the cloud. Q.E.D.

Corollary. A cloud is a *thermostat* which maintains a constant temperature and thus stabilizes the temperature of a large system of which the cloud is a part. Any heat perturbation of the large system will result in change of the cloud amount but not change of the cloud temperature.

That a system involving a phase transition can act as a thermostat is well known, e.g., ice cubes in a cold drink. The earth has two thermostats from the cloud-vapor and the water-ice phase transitions. The ice is the permanent ice of polar regions originated from the *ice ages*, not the seasonal ice of the *weather system*, thus the importance of the ice age in *climate* discussions including the greenhouse warming which is beyond a *weather* problem (and the *general circulation model*). Thus the intergalcial earth has a stable temperature, conducive to the development of life and humans. Mars and Venus have no thermostats of cloud and ice respectively and thus have no life. This point of view helps to solve the greenhouse warming as a problem unique to the ice age.

Note that phase transition from the *ordinary water* to vapor does not contribute to the thermostat effect of the earth climate because no point in the climate system has a temperature of 100°C or above. The significance of the aerosol phase is to lower the boiling point from 100°C to the dew point that actually occurs in the climate system. (The existence of thermostats on earth is due to fortuitous initial conditions. Mars and Venus are not so fortunate. Including this requirement, the probability of existence of intelligent being beyond earth is much smaller than previously estimated by Carl Sagan and others. Any efforts to communicate with extra-terrestrial beings are most likely to be futile.)

Application to the climate system. The earth surface is subject to heat perturbations of various kinds, such as the heat from the increased greenhouse gases; each will bring its own feedbacks of various kinds. In a microscopic approach to the climate study, all these must be studied in minute detail and then added together as in the general circulation model. Many mechanisms of cloud feedbacks are still not understood precisely. The grand total of many minute details is subject to large statistical error. This is the crux of the matter of the current study of the greenhouse effect. On the other hand no matter how complicated and uncertain of the detailed mechanisms involved, the thermodynamic result that the thermostats maintain a constant temperature is always true and cannot be changed. This enables us to bypass the uncertainties and complications of the conventional climate studies and reach clear cut conclusions.

Greenhouse effect. Considering the case of global warming by anthropogenic greenhouse gases, we learn from the above study that the clouds maintain a constant temperature and thus they act as a constant temperature boundary condition which will determine temperature distributions below the cloud level up to the surface of the earth. The current earth is 50% cloud covered. The clouds distribute themselves randomly and thus form a patchwork constant temperature boundary surface enclosing the entire earth. As a boundary condition, such a patchwork constant temperature surface is not much different from a uniform surface as a good approximation just as a metal meshwork is as good a shield as a solid metal surface (in a microwave oven, for example). In this way the constant temperature boundary condition determines the constant temperature of the earth between the ground and the cloud level. Therefore *there is no global warming* on the surface of the earth; the effect of the greenhouse heat addition (from carbon dioxide, etc.) is just to increase the cloud coverage, not the earth temperature.

The above thermodynamic result can be visualized from a kinetic theory point of view as follows: The anthropogenic greenhouse gases, together with their feedbacks excluding those to be introduced in the following, generate a greenhouse heat (sensible) H in the lower atmospheric and raise the temperature of the air dT. A part of the heat H_1 is used to increase evaporation of the ocean dV and converted into latent heat, which is later released at the cloud level when the increased vapor condenses to form cloud and increases the cloud coverage dC. This process simply transports greenhouse heat H_1 from the surface to the cloud level.

The cloud level is further perturbed thermally by (1) the greenhouse heat in upper atmosphere (from the cloud level up) H' and (2) the heat dissipated by the increased albedo of the increased cloud coverage dC designated by D, and (3) the conduction and convection of the remaining greenhouse heat in lower atmosphere to the cloud level, which should be exactly $H-H_1$ because under the constant temperature boundary condition of the cloud level there can be only one equilibrium temperature of the lower atmosphere, which is the original temperature before perturbation. (Thus all greenhouse heat in lower atmosphere is transported to the cloud level.) Hence the net heat perturbation at the cloud level is $H_1+H'-D+(H-H_1)$ or $H+H'-D$, that is, the total greenhouse heat $(H+H')$ less cloud dissipation, which, if not already equals zero as it should in an equilibrium state, will be absorbed by the cloud thermostat with a resulting change of cloud coverage, generating the next round of feedbacks until eventually the energy is balanced with an equilibrium dC. Therefore, the eventual heat balance is that the entire greenhouse heat $H+H'$ is removed by D, that is, by reflecting away the solar radiation by the increased cloud coverage dC.

The heat balance between the initial and final state also involves the change from the initial ocean water in the water phase to the final cloud dC in the aerosol phase in which heat of vaporization and condensation balance out. And then to the state of precipitated water (rain drops), but the latter is a phase transition of the second order with zero latent heat; hence it does not affect the energy balance. With the entire greenhouse heat removed, there can be no change of the atmospheric temperature T. Thus no global warming.

It is to be note that water balance is accomplished by evaporation and precipitation with an infinitesimal amount changed to cloud but that is enough to get rid of the entire greenhouse heat as is well known in the greenhouse studies that the greenhouse heat of carbon dioxide is much greater than the chemical heat of combustion creating the carbon dioxide.

From the kinetic theory point of view each and all of the above processes should be treated mathematically and when all are added together the thermodynamic result should be reproduced lest the latter be violated. The fact is that the processes are so numerous and so uncertain that such a treatment has so far been unsuccessful. The thermodynamic treatment is just a short cut to bypass all the complications and uncertainties.

Experimental and theoretical verification (1) It is generally agreed that earth temperature in the past century increased by 0.6°C.[1] However, there are corrections[2] due to solar irradiance, atmospheric turbidity and ENSO effects which reduce the warming to 0.18°C and includes 0°C within the error margin.[2] The main problem in recent studies has been that the expected greenhouse warming is largely not observed.[3-10] This is the well-known *missing greenhouse heat problem* and we explain it as due to energy dissipation by the increased cloud due to the greenhouse gases. (2) Cloud coverage in the past 50 years has been observed to have increased 4.1%.[11, 1, 12-15] (3). It can be shown by calculation that the above increase of cloud coverage leads to an albedo increase[16] which cancels out entirely the *clear-sky greenhouse effect* (2°C increase for doubling greenhouse gases)[17]. Thus the cloud feedback reduces the greenhouse warming to zero. (4) It can be shown by calculation that the greenhouse heat H_1 does increase evaporation dV and thus increase cloud coverage by an amount dC equal to that actually observed in (2).[18] (5) Precipitation over the past century did increase by 7.8%[21] which agrees with the amount of greenhouse heat generated in such a period. Thus all meaningful and crucial verifications are all obtained.

The final state of the greenhouse *effect* here obtained—increased greenhouse gases and cloud coverage but no increase in temperature—is a valid state because of the energy balance shown by (3) and is the legitimate succession state of the initial condition because of (4). Therefore, based on the unique theorem of differential equations, the final state here established is the only possible final state obtainable from the initial conditions. Thus there is no need to verify it by solving the complicated and treacherous differential equations of the general circulation model just as there is no need to derive the Clausius-Clapeyron equation by kinetic theory in the next century since it has already been proved by thermodynamics a century ago.

Future warming. Greenhouse warming will ultimately commence when finally the sky is 100% covered with clouds (as in Venus) so that no more cloud coverage may be generated to cool off the earth. This is expected to happen 3000 years later. However, that time is 10 times longer than the fossil fuel lifetime of 300 years, after which the carbon dioxide increase would be minuscule. Hence the eventual commencement of greenhouse warming will be moot.

Postscript. Besides the cloud thermostat the earth has already had a thermostat based on ice-water phase transition in the interaction of icesheets with ocean through iceshelves. Any heat perturbation taking place on the surface of earth, including greenhouse heat, cannot avoid being detected by this thermostat in the form of icesheet melting and sea level rising. Experimental evidence from these two indicates that this thermostat is quiescent to less than 0.1% of the greenhouse heat.[18] Since the greenhouse heat is real, the only possibility is that more than 99.9% of it has been "scooped" by another thermostat, none but a thermostat, with a shorter time scale to beat the iceshelves thermostat. The other thermostat must come from the other phase transition, water to vapor, but the transition temperature will be 100°C outside the climate system if there are only three phases of water. Since the cloud involves water-vapor transition and is a powerful heat dissipater, the indication is that the cloud is the other thermostat and the aerosol must be the forth phase of water with a "boiling point" much lower than 100°C. Thus is the genesis of the idea of the fourth phase of water. This paper is not an intellectual stunt; it merely brings out the unavoidable that should have been known already. All its conclusions could and should have been recognized earlier based solely on the established fact that the icesheets are not melting.

This work brings out the relevance and significance of phase transition equilibrium, which is generally ignored in climate studies such as in general circulation model. The reason is that phase transition involves a discontinuous function with temperature derivative equal to infinity which cannot be handled easily by differential equations and computer calculation. The temptation is just to ignore it. Thus a crucial point is missed. Super-computers cannot help. A thermodynamic theory is most suited to express the essence of the involvement of phase transition.

A similar situation has happened in the other phase transition between ice and water in the problem of ice age dynamics. The conventional treatment has led to the bifurcation of solutions of differential equation and unreal solutions.[19] However, in a realistic theory[20] of the ice involved in climate problems, the mathematical solution has demonstrated nearly infinite derivative with respect to temperature and features of neutral equilibrium, which are earmarks of phase transition equilibrium. Thus the problem can be conveniently treated by a thermodynamic theory parallel to what has been done in this paper on the greenhouse warming problem.

Addendum A (2001) on Phase transition. Modern knowledge is that when water vapor condenses into liquid water, n molecules of H_2O combine to form a cluster by *hydrogen bonds* designated by $(H_2O)_n$. The latter is a bona fide new molecule *physically*. The process is a bona fide chemical reaction involving a heat of reaction (latent heat 590 cal), the per molecule value of which is indeed the hydrogen bonding energy per H_2O molecule in the new product liquid water. But it is not recognized as a *chemical* process because (1) hydrogen bond is not ionic nor covalent bond that defines chemical combination and is not shown in the molecular formula, (2) the number n is not definite, varying in time from 1 to 10 incessantly and cannot be expressed in molecular formula [that is what liquid water is—a collection of clusters (n=1-10) of H_2O molecules that has a definite volume but not a fixed shape (shaped by container)]. Since our present concern is exclusively physical, we have no choice but to accept the phenomena in its physical reality, that is, as a chemical equilibrium process. The conventional idea of phase transition is not chemical in nature but one faking as a physical process parallel to change of state, called change of phase. It is a status half legal and half illegal comparable to a sinner in the purgatory or a concubine. Elaboration follows.

In the change of state defined in physics as represented in the equation of state all variable P, V, T are continuous and all properties of the system are differentiable with respect to P, V and T. All molecules of the system vary continuously with respect to P, V and T *en masse*. By contrast in an isothermal chemical reaction, properties of molecules do not vary *en masse* but vary *bit by bit*, then *clump by clump*, according to the heat of reaction added and vary discontinuously from one kind of molecule to another. A new *thermodynamical variable*, the molar number, appears, which vary continuously *bit by bit*. Temperature is a constant and derivative with respect to temperature is infinite. In comparison phase transition is exactly the same as the chemical reaction, but completely different from the equation of state. Trying to make a chemical reaction look like a change of state is like to take a concubine (the new molecule appeared out of nowhere) and the missing dowry is the heat of reaction (latent heat) which also disappeared. These are the source of difficulties.

A classical case of this effort is the van der Waals *equation of state* which tries to take care of the *phase transition* (from vapor to water), which is a task seemingly impossible. Miraculously it was accomplished by the magic Maxwell's rule effortlessly, like a sinner lifted to Heaven without penance. Is the Maxwell rule black magic? No! It is derived from the second law of thermodynamics. How does it work? What has been accomplished? What is left to be done?

To be brief the van der Waals equation is an equation of state for the *ideal van der Waals gas* that its molecules do not have hydrogen bonding to form clusters to form a liquid state (it is a perennial gas) and would indeed follow the S-shaped curve according to the van der Waals' equation. The mysterious, unphysical S-shaped curve originates from the fact that the van der Waals is a cubic equation of V, thus 3 roots of V. The cubic originates from the term a/V^2, the *van der Waals force* of which can be derived from the second order perturbation theory of quantum mechanics, which, alas, is valid only for large inter-molecular distances, thus is not valid for close distance near a liquid state in the region of the S curve. For *real* gases the hydrogen bonding appears and it forms molecular clusters, that is, a new compound out of a chemical reaction. Then the laws of chemical equilibrium takes over and we have an isothermal chemical reaction with constant temperature but with increasing molar number of the clusters until all molecules are included in clusters, that is, the gas is completely changed into the real liquid state. This way we have theoretically derived a process of constant temperature with latent heat changes from chemical equilibrium, which physicists facetiously called phase transition.

Even though the ideal van der Waals gas is unphysical it does not involve a Hamiltonian that violates the second law of thermodynamics (see main text) and we are legitimate to assume such a gas for theoretical purposes to derive the Maxwell rule. However, the Maxwell rule does not demand a *horizontal* straight line—any other crooked curve that produces two equal positive and negative areas out of the S curve would do. It takes an *additional assumption* to choose the horizontal straight line. When the assumption is examined experimentally, it is found that it is indeed correct as expected by Maxwell—a grand comedy of errors. The physical truth of the horizontal straight line is established by the hydrogen-bonded molecular clusters that shows that the so-called phase transition is actually a chemical reaction equilibrium process.

To clear the smoke it would be better to expunge the term phase transition entirely from physics and think of everything of it from the chemical equilibrium point of view. However, the fallacious success of the van der Waals equation gives the illusion that everything can be treated from the equation of state point of view, without regard to phase transition and latent heat. This is the origin of the doctrine to emphasize the general circulation model without due attention to phase transition in the current treatment and failure of the study of the greenhouse warming problem.

Since everything converges on the second law, the simplest way to tell school children that there is no greenhouse warming in one sentence is to quote the Le Chatlier Principle, which is a consequence of the second law thus: The increase of the extensive quantity greenhouse heat does not increase the corresponding intensive quantity temperature. (This is the correct quote to the original Le Chatlier avoiding the often misquotes and unjustified generalizations.)

Addendum B (2001) on ice ages. Ice age involves phase transition from water to ice (instead of from water to vapor) and has similarity with the greenhouse warming problem. Likewise it should be treated as a chemical equilibrium problem with discontinuous equation of state and continuous molar number as thermodynamical variable, and with discrete latent heat. It should not depend on continuous equation of state like the van der Waals equation. It should emphasize the equilibrium static status instead of the dynamic forcing change actions. But the conventional treatment of the ice age problem just follows the same track as the greenhouse problem and becomes enmeshed in the same quagmire. The continuous differential equation of such a theory runs into bifurcation, unrealistic solution, and the S-shaped curve,[19] almost a *deja vu* of what happened to the van der Waals equation.

One major difference must be pointed out in the first place. In the greenhouse problem the temperature of the earth is constant as in the phase transition and the connection of the two is more obvious. The static status stands out unmistakably. In the ice age problem the temperature of the earth is not constant and the connection with phase equilibrium is not obvious. In fact the lowering of temperature of the earth was generally considered as the major and foremost problem, to be treated as a *dynamical problem* of atmosphere science due to forcing actions of external source. And thus forgetting about the ice that is the earmark of ice age at all (not one paper on ice ages mentions the heat of fusion 80 cal), which misleads the theory astray a long way. For another, the tremendously large change of the ice mass gives the impression that it is a dynamical problem, not an equilibrium problem, which involves no change or little change that can be treated as a perturbation. The marvelous, unexpected point, as eventually emerged, is that extremely large changes can be generated by small perturbation in *neutral equilibrium*. And phase equilibrium is indeed a neutral equilibrium. Under such a condition the extremely large changes can be calculated by the perturbation theory, originally dedicated to small changes—indeed a miracle, thanks to neutral equilibrium, which was unrecognized in nearly all sciences including economics.

That ice-water phase transition takes place with constant temperature is what happens in the laboratory condition isolated from other influences. The earth is an open system, open to other influences, which may complicates the matter. For example, sun light is interacting and earth heat radiation is going out, known in atmospheric science as albedo and infrared feedbacks, both affect the heat balance of the ice age glacier system and the equilibrium status.

On the other hand chemical equilibrium, defined as $dS=0$ where S is the *entropy*, may be considered as a special case of a more general kind of equilibrium that may include mechanical equilibrium, defined as $dV=0$ where V is the potential energy. $dV=0$ is related to *neutral equilibrium*, of which phase equilibrium is also a kind. $dV<0$ and $dV>0$ are related to *unstable and stable equilibrium*, of which $dS>0$ and $dS<0$ are also a kind.

From the point of view of *general equilibrium*, the ice age glacier system may be considered as in a *neutral equilibrium state* which can be proved thus: The glacier does not move in one day, albeit it moves infinitesimally—taking 50,000 years to complete the journey. It is certainly not one in stable equilibrium, which does not move forever. It certainly is not unstable equilibrium which collapses instantly. Therefore the equilibrium of the global ice age glacier system must be the last alternative—the neutral equilibrium.

Now the *general ice age system* includes the ice-water phase system as a part, which is already in neutral equilibrium. Phase equilibrium is already studied in the greenhouse problem. The only thing new now is to establish the neutral equilibrium condition of the other factors in the open system beyond the ice-water system, that is, the ice-albedo and infrared feedbacks, then the whole ice age system will be in neutral equilibrium, which is what we want to see as demonstrated by empirical evidence, and which would establish the principles that control the equilibrium status of the global ice age climate system. This task is to be carried out as follows.

This system is a *neutral equilibrium system under perturbation* and can be treated by the *perturbation theory*, which is an approximation of the exact dynamical theory for simplifying the mathematical treatment as one dealing with a small perturbation but without missing the spirit of dynamic theory. The theory has been used by Laplace to calculate the advance of perihelion of the planets and by modern theoretical physicists in 90% of the quantum mechanical calculations of atomic, molecular, nuclear and particle physics, a staple theoretical tool, now for geophysicists to catch up.

The perturbation theory concerns the changes of an already solved *unperturbed problem* by the addition of a small *perturbation*. The simple Newtonian solution of planetary orbit which sophomores learn in one hour is that the planetary orbit is a perfect ellipse (due to the gravity of the sun only). This may be considered as an unperturbed problem because there are also gravity forces from the other 8 plants of the solar system, which are much smaller and can rightly be neglected as Newton did. But they nevertheless produce small but measurable deviations from the Newtonian solution, one of them is the advance of perihelion. Since it was beyond the original Newtonian solution, it was once thought to be beyond science and thus is the proof of God's free will. Then came Laplace who by using the perturbation theory has shown that it is due to the small gravity forces of the other 8 planets combined. Thus the complete *perturbed problem* is solved by a perturbation theory. Since then solar system is known to be a purely mechanical system—no more than a clockwork, and out of God's hand. Mathematically the result of the perturbation is to change the Newtonian constants of integration into slowly changing variables, a change that will reappear in the perturbation theory of the ice ages. Thus the perfect ellipse changes into a rotating ellipse, observed by astronomers as the advance of perihelion.

The ice age problem starts with an unperturbed problem of the phase equilibrium of ice and water with constant ice mass M and equilibrium temperature T of the earth. Then "enters" the Milankovith perturbation F_{ex} (like the Laplace perturbation) with the albedo F_{al} and infrared F_{in} feedbacks and we have a new perturbed problem to solve by the perturbation theory. Now the perturbed state is still in neutral equilibrium and thus the three F's balance themselves out to zero. This is now the basic dynamical principle of the perturbed problem.

It so happens that, as time goes on, the feedbacks, unexpectedly, are much greater than the Milankovitch forcing by 30 times, a fact that can be recognized from empirical facts by freshmen but overlooked by pundits. It gives the misleading impression that the Milankovitch force is insignificant, missing the fact that it is instrumental to get the whole process started, a point which is made clear only after the whole perturbation theory has been solved completely. Thus by the spirit of perturbation theory we can neglect F_{ex} for the purpose of setting $F_{al}=F_{in}$ for perturbation calculations (but not for all other occasions). Now these two feedbacks are known to be in balance to maintain the heat balance of the earth—the earth cools itself down to balance its heat budget due to the addition of the albedo feedback. So the introduction of this equality condition is quite natural in the spirit of the perturbation theory to neglect F_{ex} in the unperturbed solution. This equation $F_{al}=F_{in}$ will be the physical principle used in the perturbation theory, which appears in Eq. (3) of paper (b) in the earlier work. The system is thus in neutral equilibrium expressed mathematically as we want to establish.

The *dynamical variables* of the *global ice age climate system* are $M(t)$, the ice mass and $T(t)$, the global temperature of the earth. They are the *dependent variables* to be solved by a set of differential equations determined by the dynamical laws by the perturbation theory. The equation determining the change of $M(t)$ is Eq. (1) and that for $T(t)$ is Eq. (2). Equation (4) is that of global energy balance of the three F's. F_{al} is given by Eq.(6). F_{in} by Eq. (7). F_{ex} is the Milankovitch forcing. The equations involve the surface and volume of the icesheet and a geometric relation is needed to connect surface and volume, which is given by Eq. (8). This set of simultaneous integral-differential equations (1-8) form a complete set that determines a *unique solution* of the problem. The results are displayed graphically in the paper. The conclusion of the increasing of icesheet mass and the lowering of global temperature of an ice age period so obtained are compared well with experimental results, fulfilling the objectives of a dynamical theory now coming out of a perturbation theory.

The previously first *published* work, paper (a), though formulated differently because of historical conditions, can now be identified in paper (b) as a perturbation theory to determine the changes of the equilibrium condition of a system in neutral equilibrium under an external perturbation. The changes are indeed large compared with typical perturbation theory because of the neutral equilibrium condition.

In such a treatment the neutral equilibrium of phase transition of water to ice can now be identified as a chemical equilibrium in which the mass M is actually the molar number of the ice phase, its dynamical equation being Eq. (1), and the temperature is the chemical equilibrium temperature, both are the *dynamical variables specifying the chemical equilibrium system*, now perturbed by the external forcing and its feedbacks. The heat of fusion is actually the heat of reaction of the chemical equilibrium. The identification of the problem to a chemical equilibrium is thus complete.

That a weak Milankovitch forcing can generate a large glacial advance is mainly due to the neutral equilibrium condition, like a sphere rolling a long way on a plane surface without little external assistance. A conventional dynamical theory seldom leads to a neutral equilibrium condition and cannot be in a position to explain it.

Equation (1) on ice mass change has been praised as a major contribution because heat of fusion has never appeared in past studies, so much so that, in 1982, in paper (a), which I first published on the problem, the Editor changed the title to "Latent heat of fusion and its importance for glaciation cycles," much to my displeasure. That was a time the idea of neutral equilibrium has not yet been introduced into the study and Eq.(4) of paper (a) was introduced as an empirical parameter to play a role that was later to be replaced by the neutral equilibrium condition. Even though its correctness as a physical fact is beyond doubt but it aroused the logical suspicion of introducing the conclusion of the theory as the assumption to start the theory, which is tautology. This charge was unfair, because I introduced only one empirical number and have harvested infinitely many new numbers, which is indeed a good bargain, not tautology.

Another major criticism was that the derivative with respect to temperature is infinite. Now we know this is due to the reconstruction of the van der Waals equation by the Maxwell's rule to make it realistic. It is indeed infinite in the reconstructed van der Waals equation. In spite of these severe criticisms the Editor Steven Schneider kindly published the paper in probation, with an Editor's Note in the bottom of the first page contending that the paper is nevertheless published for the purpose of spurring debate on the large remaining uncertainties over the cause of Pleistocene glacial cycles. In the 20 years thereafter I have heard no debates. But I have new ideas, about one a year, to piece the jigsaw puzzle together to form a complete, all encompassing theory. The past work is summarized in papers (a, b, c). The completion of the jigsaw puzzle is the best proof of the theory, because there is only one way to fit all parts together neatly. A jigsaw puzzle is two-dimensional and cannot be rendered into a one-dimensional logical presentation of the written language in a satisfactory and complete manner. Here we take a novel approach to achieve this end, which continues to the next Addendum.

The large remaining problems appearing in the jigsaw puzzle are much more than Schneider had in mind or anyone can possibly imagine.

They include the following: Is carbon dioxide level in atmosphere the cause or effect of the ice ages? If the former, then how to incorporate it in the dynamical theory? What is the role of cloud in ice age cycles and how to incorporate it in the dynamical theory? Why there is no ice ages before Pleistocene? This question eliminates 90% of all the existing and future theories. Once glacial advance has started why does it not go all the way to the equator as many theories of glacial advance have predicted it as a matter of course? The current Pleistocene glacial cycles are known to have been in the past 2 million years or so, a minute fraction of the geological history. What was the earth doing for the rest of the past times? There are indications of glaciations 300 million and 600 million years ago. What happened then? Is there a grand cycle of 300 millions of *ice age groups* in history? And what is the cause of the 300 million year cycle? Those glaciation periods are times of high carbon dioxide level which are supposed to cause extreme greenhouse warming. Then why the extreme cold glaciations? A glacial cycle of 100,000 years involves drastic climate changes. But why there is an at least 12000 years of extremely stable and temperate climate in the interglacial period, in which we find ourselves today. This is related to the problem why is the next ice age long over due? After we have overcome the nightmare of greenhouse warming, how do we prepare ourselves for the peril of the cold of the next ice age in which all the Yankee land will be buried under mile high ice mountains? The 2 million years of *harsh* climate of ice ages coincided with the emergence of humans (Homo Sapiens Sapiens) from Homo Sapiens, and the 12000 years of *mild* climate coincided with the sprouting of human civilization. Are they just coincidences? And why are they contradictory in climate? Are there ice ages, humans and civilizations on other planets in the solar system and other galaxies? A complete ice age theory will be in a position to answer these questions and to fit the jigsaw pieces together.

Addendum C (2001) on a canonical theory of climate on the ten million year time scale. To wrap up our study of global climate it is worth to note that the theory of greenhouse warming can be worked out as a special case of the theory of ice ages because they are both general theory of the global climate, the only difference being one concerns the cloud-vapor and the other concerns the ice-water phase transition. The ice age theory is more general where the global temperature is a variable. The greenhouse warming theory is special in which the temperature is a constant. This fact we have established by phase equilibrium in the cloud and the fact that the cloud layer of the earth, the *cryosphere*, forms a constant temperature boundary condition that determines the constancy of temperature of the earth surface.

This leads to the following simplifying conditions for the greenhouse problem.: (1) the polar icesheets do not melt and sea level does not rise as a result of climate change (sea level does change as a result of geological changes as already discussed). (2) The ice thermostat is out of function here and the cloud thermostat is the only one operating. (3) The infrared feedback is now zero, having no part to play. The albedo feedback is that of the cloud, which can be experimentally determined by the ERBE Satellites. (4) The ocean no longer plays the role of a *heat* reservoir as in the ice age theory but now plays the role of a constant *water* reservoir— global evaporation M equals precipitation M_1 plus new clouds formed M_2. Here are two new dynamical variables: the mass M_1 of water from precipitation that returns to ocean and the mass M_2 of water of the cloud that remains in the sky and generate albedo feedback. The heat of fusion of the ice age theory is now replaced by heat of vaporization.

Now we can set up differential equations for the perturbation study of the greenhouse problem in a way parallel to the ice age theory to solve for the dynamical variables M_1 and M_2, the other dynamical variable T temperature being already determined to be constant. Fortunately the detailed mathematics can be short cut by the knowledge already well known in greenhouse warming studies that the greenhouse heat generated by carbon dioxide is much greater than the heat of combustion of fossil fuels that produce that carbon dioxide so that the latter is always neglected in the total heat budget of the greenhouse warming problem. This means the mass M_2 can be neglected when compared with M_1 (but not in other places), and therefore the latter can be approximated by M. A detailed set up and solution of the differential equations analogous to the ice age theory can always be done and the results will not be different from what we have obtain from the short cut here. Thus it will not be done here.

The dynamical problem can now be solved without solving differential equations by computer as in the ice age case as follows: The forcing term of the equation system acting as the perturbation is the greenhouse heat generated by fossil fuels. This heat is added to the unperturbed system in neutral equilibrium with constant temperature and we are interested in finding out the perturbed results. Since the temperature is constant, the perturbing heat cannot go into any sensible heats that would raise temperature. Therefore it can go only into latent heat and the only latent heat available is heat of vaporization Thus all greenhouse heat is used up to generate vapor and is recovered *in total* during condensation in the cloud level. Two things now follow: (1) By global heat balance this heat from condensation must be got rid of the earth by albedo feedback of the new cloud formed that floats in the sky. Therefore the latter, which is M_2, can be calculated by the conversion ratio given by the ERBE satellites. (2) The other part M_1 that falls off sky can be approximated by M which can be calculated by the total greenhouse heat divided by the heat of vaporization.

The results of the calculation of the theoretical values of the dependent variables M_1 and M_2 of the dynamical equations so obtained can be compared with the experimental values of cloud increase and precipitation increase observed in the past century to conclude the dynamical study. Good agreement is obtained as reported in the other papers. To be specific, the theory predicts correctly *all greenhouse effects*, that is, effects on precipitation, cloudiness and warming. That T is predicted to be a constant means there is no greenhouse warming.

On the other hand the general circulation model deals with the mechanical energy and properties of a fluid and is not equipped to deal with the chemical properties of the molar numbers and latent heat and therefore ends up with delinquent performances in the prediction of M_1 and M_2 and miserable mixed-up in the assertion of greenhouse warming coming out of T.

*   *   *   *   *   *   *   *   *   *   *

This is not an intellectual game for the entertainment of the leisure class; it is as serious as a fight for life or death and the survival of the species. It shows all climate problems in the current few million years (of the ice age epoch) can be treated in a *canonical* approach that is systematic, reliable, and controllable. The canonical approach is to treat the system basically as an equilibrium system to start with and all changes taking place are to be treated by the perturbation theory to find out the deviations from equilibrium, not by a full-fledged dynamical theory for the complete dynamical history as in the general circulation model.

In Addendum B we have summarized the previous work on the ice age theory in papers (a,b,c) and add new insight from later emerged ideas including phase transition, chemical equilibrium, neutral equilibrium and perturbation theory. In the above (Addendum C) we have shown that the previous greenhouse warming study can be made as a special case of the ice age theory. The suggestion is that the two can be combined into one general theory which not only includes the two as special cases but can also be applied to other similar problems in a standard approach that may be considered as a *canonical theory* dealing with *canonical dynamical variables* such as M_1, M_2 and T. Such a theory is supposed to be fundamental and cannot avoid the basic problem in synthesizing two special theories concerning the relation of basic principles and historical evolution,a special case of the general problem of unification of philosophy and history.

A general guide in physics is provided by Professor Enrico Fermi's famous dictum that short term rapid changes are to be treated by the (full fledged) dynamical theory and long term slow change by the statistical (equilibrium) theory. It benefited me throughout my life since my world renown Ph. D. Dissertation on nuclear fission up to the ice age and greenhouse research. These problems are all long term slow change ones suitable for the equilibrium approach and the dynamical approach is out of the place according to the dictum. The canonical theory we are looking forward to is thus expected to be inclined to the equilibrium approach.

There are indeed many problems similar to ice age and greenhouse effect. In fact they include all climate problems in the recent millions of years (the ice age epoch), in which the two thermostats dominate the world climate and thus these problems follow a similar dynamical history as the two already studied. So we have a category of phenomena following a similar evolutional pattern and controlling principles. Their studies naturally form a self contained branch of science. Like elastic body dynamics dealing with the dynamical behavior of all elastic (in contrast to rigid and fluid) bodies following an exclusive set of laws separate from others (rigid and fluid bodies), thus forming a separate branch of science.

Therefore we can look forward, in a similar way, to establish a separate, independent branch of geophysical science called *Pleistocene climatology*. Its phenomena category encompasses the climate changes in current millions of years (separate from the ice-free planetary climate of the past billions of years and the short term weather related climate changes). Its dominant feature is the equilibrium state established by the two thermostats, which appears as the unperturbed state. Various kinds of perturbation may occur. All climate changes are thus treatable by the perturbation theory.

These ideas may be made more transparent by a comparison with the dynamics of elastic bodies. In the latter the category of phenomena encompasses those of all elastic bodies separate from rigid and fluid bodies. The external-force-free state is the equilibrium state. The external forcing may cause motions, essentially vibrations, which can be treated as a perturbation of the equilibrium state.

Every branch of mechanics is formulated by an equation of motion which is one and the same based on the invariant laws of nature which includes (1) the *universal* mechanical laws, that is the Newton's laws and (2) the *branch* mechanical laws pertaining to the *particular branch of mechanics*, for example, (a) Hooke's law for elastic bodies and its mathematical formulation in Euler's equations of motion, and (b) the phase equilibrium law for the thermostats of the climate system.

Besides the universal and invariant parts (1,2) the equation of motion also contains a part (3), an external force term representing the *parochial* laws of nature specific to the particular case, and is *different from case to case*, such as electric and gravitational forces in the solid body system and the greenhouse forcing and Milankovitch forcing in the climate system.

In solving any mechanical problem, first is to identify which branch of science it is in, to choose the correct *branch* equation of motion, which ordinarily is obvious—one never mistakes a rigid body for an elastic body. Then to nail down the specific forcing for the specific problem concerned. This is usually the crucial part and most prone to mistake. And a test of the mettle of the scientist. Once this is done the rest is the solution of the differential equation of the dynamic system. Today this can always be done by computer and usually is no longer a serious scientific concern.

This leads to the outline of the future branch of science that is the Pleistocene climatology. Indeed the theories of greenhouse effect and the ice ages as recently developed fit the profile of the general theory closely. We endeavor to achieve this goal. In such a general science once the equation of motion is established the correct forcing will lead to the correct results without further question. Historical development of science is usually not that simple. Not only the forcing could be mis-directed, even the equation of motion could be wrongly chosen. The general circulation model of climate studies uses a fluid mechanics model whereas the climate state is obviously a *chemical* equilibrium state, not fitting in a fluid motion model. The historical difficulty may be explicated in the discussion below.

Newton's theory that explains the planetary system marvelously made use of Newton's equation of motion and Newton's gravity force at the same time. Logically one experiment cannot prove two independent assumptions involved at the same time. It is only the hundreds of successful applications that made every sensible person believe that both are correct in spite of the logical difficulty. We are just beginning to establish the Pleistocene climatology and the only convincing proof for both the equation of motion and the forcing is the large multitude of successful applications that prove the theory as a whole.

Before doing so a review of the history is helpful. The planetary system problem was first solved by using the theory of Newton, which deals with the unperturbed problem of the gravity of the sun only. Then the minor gravity forces from the other 8 planets were treated as perturbation to find out the minor changes of the Newtonian orbit, such as the advance of perihelion mentioned earlier. In doing so the constants of the unpertubed solution become slowly-varying variables, showing the small changes of the orbit, such as the advance of perihelion. For the greenhouse problem the unperturbed M_1 and M_2 do change slowly in response to perturbation and can be calculated. The change of T happens to be zero as a special case. In the ice age problem the changes of both M and T can be calculated. For other problems in the current millions of years, the same can be done by the same methodology.

The Newtonian theory's successes of his time are enough to establish its truth. Any further successes, such as the advance of perihelion and the space ship flights are un-needed embellishment for the established glory. In the climate studies, the ice age theory developed here has already had enough successes to support itself. Any further successes as applied to the list of problem three pages ago would enhance its credibility, approaching the goal of an established science. The solutions of those problems are:

The carbon dioxide level changes in an ice age is the effect, not the cause, of ice ages. The cloud effect needs not be included in the differential equations because there was no cloud in the time of glaciation. No ice ages before Pleistocene because the Antarctica continent has not reached the polar position by continental drift. The glacial advance stopped at 40° N latitude because the icesheet is 3-dimensional, not 2-. The glaciation 300 and 600 million years ago are due to other continents having drifted to a polar position; those glaciations are indeed in other continents now not in polar position. The high carbon dioxide level of those glacial period is of geological origin of rock erosion and its greenhouse effect is irrelevant to glaciation. The stable temperature of the 12000 years of interglacial (the time between two glaciated periods of clear sky from the tail of one to the head of the next), is the time it takes for the sky to change from clear to fully clouded 6000 years, plus, after the reversing of ice age cycle, from the fully clouded to half clouded (the present day time) 3000 years plus another 3000 years to change to the clear sky to start the next glaciation. The numbers add up to 12000 years that is the interglacial time.

This is also the reason why the next ice age is long over due and has still 3000 more years to wait (very fortunate for those living in the Yankee land—the ice margin of the ice age coincided with the Mason-Dixon line). When the next ice age does come 3000 years later the way to save the Yankee land from snowed under is to get help from the greenhouse effects; the current greenhouse forcing is of the same order of magnitude of the Milankovitch forcing. Unfortunate for us, fossil fuels are finished 2700 years ahead of time. But we can produce carbon dioxide by other sources such as by baking sedimentation rocks (Chinese 2000 years ago did bake rocks 15 years to make a tunnel of 60 ft through mountain to divert river water in the world-famous ancient hydraulic engineering project Tu-Chiang-Yien) with heat from nuclear energy, which by itself cannot stop the ice age but can by the carbon dioxide coming out of calcium carbonate through the greenhouse effect, a fact we have learned previously.

Ice age *alternative climate changes* cause the expansion and retreat of the *forests*, which promoted the evolution of Homo Sapiens from quadruped to bi-ped locomotion, freeing the fore limbs to evolve into hands. This is then followed by the development of tools, technology, complex social structure, oral language, and highly developed brain ending in Homo Sapiens Sapiens. The *interglacial stable and mild climate* then promoted the development of agriculture, the sedentary society, the written language, science, arts and civilization.

Thus we are in an age of wisdom, belief, light, and hope, but also in an age of foolishness, incredulity, darkness and despair. We are going straight to Heaven and to hell. We can create everything; we can destroy ourselves. We have created the double chastity belts originated from the dread of greenhouse warming and nuclear hazard which can choke off the energy and vitality of the industrial civilization and extinguish the light of humanity. By the law of statistics humans will become extinct because once the fluctuation passes the threshold and the flame goes out, no fluctuation can bring it back again. Is here no hope in the future? Yes! There is— 300 million years later when the curtain will be raised for another drama of evolution with equally splendid performances.

Enough for things on the earth. What about the outer space? The chance of a temperate climate planet with resources for life in the galaxy has been estimated to be in the millions. But adding the condition of the ice age of the right magnitude and the right frequency, the probability is reduced to one in one galaxy. And we have preempted it all. By the same reason there will be intelligent beings, one species in one galaxy outside the Milky Way at an auspicious time somewhere in the stellar universe. Communicate with them? It would take thousands of years for the radio signal to reach there and same time to get reply. We cannot wait that long.

While the above answers to the problems raised three pages ago are largely related to the Milankovitch forcing, this needs not be so in general. Once the forcings are identified, the canonical approach can work out the answers. In the past century world temperature did fluctuate $\pm 0.5°C$ in the decadal time scale. In the Little Ice Age of Europe temperature did drop for hundreds of years. In the past 5 to 10 thousand years the continents are drying up resulting in arid central Asia and desert Sahara. It needs only to identify the proper forcings, then their consequences can be predicted from the canonical theory. The Heinrich events are global climate changes involving glacial ice and temperature changes with stochastic occurrences and may be explained by the forcing of comet impacts, which are discontinuous and stochastic. The origin of Younger Dryas event is not known but it should be explainable once the forcing is identified. The El Niño phenomena fall into the category of millennial climatology and can thus be treated as the *perturbation of the neutral equilibrium state* (see main text). Without the introduction of the new branch of science, it is likely to be considered as a *dynamical problem* of the interaction of two fluid systems, the ocean and the atmosphere, as did by Philander. On the other hand the climate change causing the extinction of dinosaurs cannot be explicated by the canonical theory because it is beyond the ice age epoch.

Such a list of impressive accomplishments of the ice age theory is a credible support to the veracity of the proposed Pleistocine climatology. Its establishment as a branch of Newtonian science in history is as interesting and fascinating as that of the Newtonian theory itself—to unravel the logical problem of proving a theory that has two assumptions, that is, the equation of motion and the law of force. It was the meticulous search all over the universe among three dozen candidates from the bottom of deep ocean to the interstellar space for a forcing of a period of 10^5 years, which ends at the Milankovitch forcing as the only qualified one that can provide the heat perturbation of 2.8×10^5 cal needed as heat of fusion to generate an ice age. That settles the one assumption on the choice of forcing (as the simplicity and universality of the gravity force did for Newton) and the other on the equation of motion can be settled by experiment conclusively.

The establishment of the Pleistocine climatology shows that the greenhouse effect theory that calls for no warming is a special case of it and thus can present itself to claim its birth right of inheritance excluding all impostors. An alternate theory claiming the coming of global warming is as impossible as an alternate way to piece together a jigsaw puzzle and as preposterous as to explain the planetary system by the heliocentric theory, and as quaint and as moot as Don Quixote's antics and aspirations.

Innocent but deplorable; without malice but with great calamity.

References

1. J. T. Houghton, Ed., *Climate Change: The IPCC Scientific Assessment.* Cambridge University Press, Cambridge, 1990.

2. Z. X. Wu, R. E. Newell and J, Hsiung, JGR, **95**, No. D8, 11799 (1990).

3. J. E. Hansen and S. Lebedeff, JGR **92,** 13345 (1897).

4. T. R. Karl, H. F. Daiz and G. Kukla, J. Clim., **1,** 1099 (1988).

5.T. R. Karl and P. D. Jones, Bull. Am. Meteorol. Soc.,**70**, 265 (1989).

6.R. C. Balling, Jr. and S B Idso, JGR **94**, 3359 (1989).

7.K.Hanson, G. A.Maul and T. R.Karl, Geophys. Res. Lett. **16**, 49 (1989).

8.R. W. Spencer and J. R. Christy, Science **247**, 1558 (1990).

9.IPCC, Scientific Assessment of Climate Change, UN, Geneva, 1990.

10. L. D. Kahl, D. J. Charlevoix, N. A. Zaitseva, R. C. Schnell and M. C. Serreze, Nature, **361**, 335 (1993).

11. S. G. Warren, C. J. Hahn, J. London, R. M. Chervin and R. L. Jenne, U. S. Department of Energy Publication DOE/ER-0406, Wash. 1988.

12.A. Henderson-Sellers, Global and Planetary Change **1**, 175 (1989).

13. J. K. Angell, J. Climate **3**, 296 (1990).

14. G. R. Weber, Theoret. Appl. Climatology **41**, 1 (1990).

15. C. K. Folland, T. R. Karl, and K. Ya. Vinnikov, IPCC, Scientific Assessment of Climate Change, Vol. 1, 185, 1990, UN, Geneva.

16. V. Ramanathan, B. R. Barkstrom, and E. F. Harrison, Physics Today, **42**, 22 (1989).

17. R. D. Cess, G. L. Potter, J. P. Blanchet, G. J. Boer, A. D. Del Geneo, M. Deque, W. L. Gates, S. J. Ghan, J. Y. Kiehl, A. A. Lacis, H. Le Treut, Z. X. Li, X. Z. Liang, B. J. McAvaney, V. P. Meleshko, J. F. B. Mitchell, J. J. Morcrette, D. A. Randall, L. Rikus, R. Roeckner, J. F. Royer, U. Schlese, D. A. Scheinin, A. Slingo, A. P. Sokolov, K. E. Tayloy, W. M. Washington, R. T. Wetherald, and I Yagai, JGR (in press).

18. Peter Fong, Bull. Am. Phys. Soc. **35**, 2356 (1990) and to be published.

19. M.Ghil, in Climate and Geo-Science, A. Berger et al. Eds., 211-240. Kluwer Academic Publishers, (1989).

20. Peter Fong, Climatic Change, **4**, 199 (1892).

21.*Trend '93*, Boden et al, Eds.,Oak Ridge Nat. Lab. (1994). pp. 765-984.

Part II Research Papers on Nuclear Hazards

The ultimate direct measurement of low level radiation effects

Long life expectancy as the beneficial effect of low radiation

Low level radiation extends life span by cutting death rates of disease including heart disease, cancer, stroke and so on

Reexamining nuclear energy safety

The above is a part of the research work on energy and environment published in book (by Macmillan Publishing Co.), in national, and local newspapers and magazines, in invited papers to American Physical Society, and delivered before public lectures, including the inaugural of the Professor K. L. Cheng Distinguished Lecture Series, UMKC.

The Ultimate Direct Measurement Of
Low Level Radiation Health Effect

Peter Fong

Physics Department, Emory University

Atlanta, Georgia 30322

Abstract

The nuclear weapons tests above ground from 1952 to 1978 generated an additional radiation of 30 mrem/year in the atmosphere which, according to U. S. vital statistics published, induced a reduction of 418,000 cancer deaths in this period (7.3% reduction yearly). This indicates the beneficial effect of low level radiation, and casts doubt on the assumption of linear extrapolation without threshold. This "experiment" is equivalent to the explosion of 50 Chernobyl type reactors and thus its result implies that the entire nuclear power industry of the world is absolutely as safe as the world of 1952 to 1978 we have lived through.

Moreover, this "serendipitous experiment" is a controlled experiment with space and time elements under manual control and with the entire population of the United States without a single exception exposed to substantial amount of nuclear radiation without their knowing and consent to guarantee absolutely neutral experimental results without using placebo controls. The experimental results obtained is complete, wholesome, not reproducible under all circumstances and thus is priceless. They should be considered as the ultimate information on biological effects of radiation on humans overriding any other source of information.

Introduction

The atomic bomb experience 56 years ago indicated that high dose of radiation may be lethal and low dose may cause delayed cancer. The dose-effect relation is linear but no accurate information is available when the dose is near zero because of low statistics. Nevertheless the established linear line seems to extrapolate to the origin (Figure 1). Thus arose naturally the assumption of linear extrapolation without threshold (LEWT), which has been the foundation of all regulations concerning nuclear safety.

However, this assumption is not automatically true experimentally. As far back as 100 years ago W. Schrader found that radiation (X-rays) may give guinea pigs immunity to diphtheria, which is lethal otherwise. This was the first discovery of radiation hormesis (beneficial effect). Similar experiment has been done on fruit fly. A large body of literature has been accumulated and reviewed.[1,2] (Of course even a single photon might cause DNA damage leading to cancer death. Both adverse and beneficial effects could happen. It is the net effect of the two that is relevant. The net effect is understood when any effect is referred to in this paper. Also in this paper we are concerned with the long term effect of nuclear radiation only, not the short term effect which is not controversial.)

The notion of radiation hormesis implies that the extrapolated straight line does not go straight to zero at the origin. That it must go to zero at origin is demanded by logic—no cause, no effect. But before it goes to zero, unusual things may happen. Radiation homesis means the extrapolated curve may go below the straight line, representing a smaller risk, and, in the extreme case may go even below the abscissa, forming a dip (Figure 1), before rising up to zero at origin as demanded by logic. The point of the dip represent a condition that radiation is *beneficial, not harmful*, a notion that flies at the face of common sense, but has shown to exist, such as by Schrader and many others and is reasonable from the evolutionary point of view (details later) For many years, "to dip or not to dip, that is the question." We will refer to *the dip* defined here from time to time.

Half a century after the original study of nuclear bomb effects, new studies were made indicating radiation hormesis.[3] The 120,000 survivors of nuclear bombs did not die off *en masse* from cancer. Some have even achieved longevity (immune to all diseases) with good statistics.[3] It is one case that proves the existence of the dip.

This proof of radiation hormesis surprised many but did not convince everybody. It has been challenged by an alternative explanation. The long survivors might be just a sturdy breed, born resistant to diseases. That might be the reason why they survived 56 years ago and that is the reason why they are surviving now, disregarding nuclear radiation (*Time*, June 23, 1997). This possibility is based on an unproved hypothesis but is a plausible one and may apply to many experiments of this kind on radiation hormesis. The logical problem is that the population sample selected for study may be biased in favor of a special trait, which happens to be the point to be proved by the experiment, thus committing tautology.

A conclusive experiment on radiation hormesis thus must be done on a large population sample, not just to improve the statistics, but to avoid unwitting biased conditions in sample selection. The ideal experiment is one using the entire population as the sample, which would be the ultimate.

HYPOTHETICAL EFFECTS OF LOW-LEVEL RADIATION

ON THREE ASSUMPTIONS

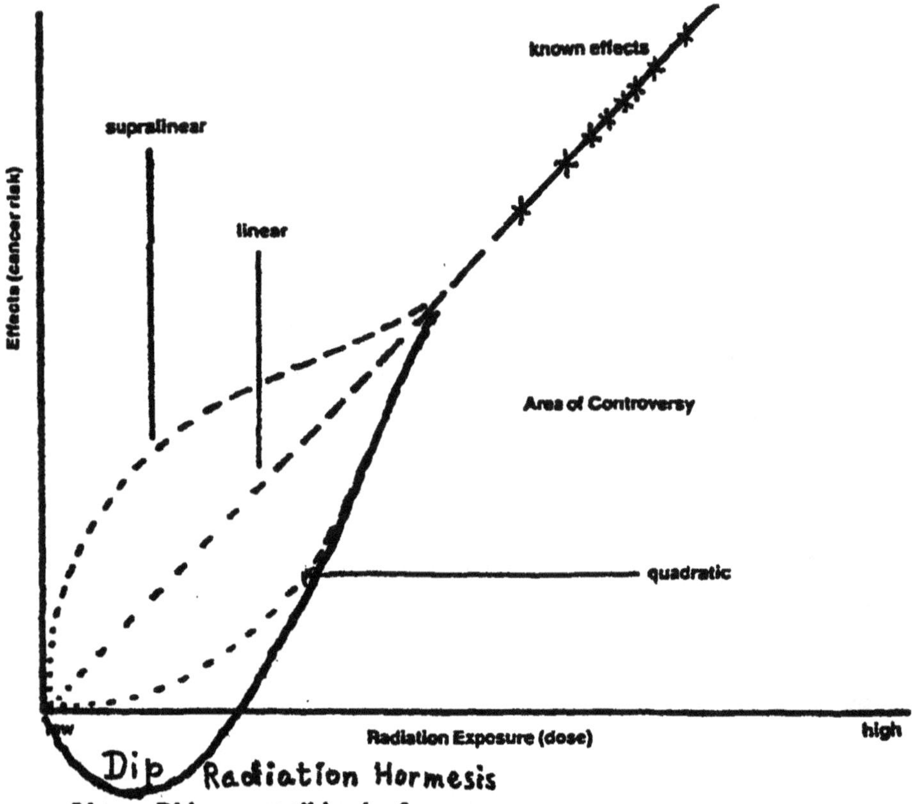

Linear: Risk same at all levels of exposure

Quadratic: Risk less at low level because cells repair themselves

Supralinear: Risk greater at low level because damaged cells survive

Source: Michio Kaku and Jennifer Trainer, eds., *Nuclear Power: Both Sides* (New York/London: W. W. Norton, 1982), p. 30.

Figure 1

A nuclear experiment involving the entire population of the U.S. to be exposed to a large amount of low level radiation from nuclear fission chain reactions (bomb or nuclear explosions from power plant or waste dump) is impossible and unthinkable from the practical, economic, political and moral points of view. But it does not prevent one from happening serendipitously. When it did happen, it would be the first and ultimate direct measurement of low level radiation effects. Its results would be the most important information over-riding all other indirect experiments with small samples which are at the risk of skewed sample selection and with miserably poor statistics.

The Serendipitous Ultimate Experiment

And one serendipitous experiment of this kind has indeed been actually taking place. It was the nuclear weapons tests in air in the two decades of 1950s and 1960s, which generated nuclear radiation from the explosion of 500 nuclear bombs, with an increase of radiation level of the atmosphere equal to 30 mrem/year during the peak test years[4] (30% of natural background radiation), a level typical of nuclear contingencies and will be the proper level to study the most relevant radiation effects including the basic issue of radiation hormesis. And indeed the entire population of U.S. was exposed, resolving the difficulty of sampling the population for experiment. No other experiments have better statistics.

The weapons tests will give us the real-life simulation of the ultimate experiment we desire. Comparing it with a nuclear explosion such as Chernobyl, the generation and composition of the source of radiation (radionucleides) are exactly the same. The transport and disperse of this source in troposphere and stratosphere are exactly the same, especially the long term effect, which is the main concern of nuclear issues whereas the short term acute effect is not a point of contention. The effects of this radiation on humans is exactly the same. The weapons tests are thus the ideal experiment we desire.

A nice feature of this "experiment" is that it covers a time period of 26 years. Thus it takes care of most of the delayed cancers in 30 years, which is a complicating factor in the calculation. Moreover, the experiment so performed is a *blind* one, without reference whatsoever to adverse or beneficial effects or to cohorts characteristics. Thus there is no need to do a costly, cumbersome and distasteful *placebo* experiment to assure objectivity, saving billions of research dollars. When the result of the experiment is revealed (yes of no for radiation hormesis) it cannot be challenged by anyone on any basis whatsoever.

This ultimate experiment was serendipitous since its manifested purpose was to test nuclear weapons. This purpose was so overwhelming that the possible uses of the results for all other purposes were completely overlooked. A great opportunity to study radiation hormesis deliberately and systematically was missed. The nuclear safety issues are still debated inconclusively to this day.

Moreover, because of serendipity, even though the hardware part of the experiment has been done, the resulting numerical data were not taken deliberately.   Fortunately it was also accomplished by others serendipitously as a routine part of the national health program to collect cancer statistics[5] yearly. And they provide us unwittingly and profusely 56 years of statistics adequate for our study. Still, because of serendipity, the raw data sat on library shelves unnoticed by nuclear scientists for 56 years even though the conclusion on radiation hormesis could have been drawn from this collection of data in a mere hour.

Finally the one-hour job was done again serendipitously for the third time. It was in connection with the work to prove that the altitude dependence of cancer mortality rates of the U.S. is due to radiation hormesis, not due to any other factors causing cancer (to be reported separately). To do this all cancer factors were considered when we study cancer statistics including a plot of cancer mortality rates[5] as a function of time in the past 56 years (now Figure 2 here). This led to the discovery of this paper. It was like the discovery of a mother lode accidentally.

As a routine, a smooth curve is to be drawn over the data points. After some struggle it was found that a good smooth curve could be drawn for all the data points before 1952, another for after 1978 and these two join smoothly together into a single smooth curve shown in the dashed line in Figure 2, indicating a general slowly increasing trend of the cancer death rate over the past half century, but with an outstanding exception for the years from 1952 to 1978, which form a smooth curve by themselves but fall below the half-century smooth curve systematically. See Figure 2.

Faced with such a comprehensive evidence, an experienced detective would suspect that the world was peaceful but some outlaws had intruded in 1952, making havoc for two dozen years, and then left leaving only the teeth mark of a bite on a round cookie in Figure 2. The space-time elements of the teeth mark are the breaking point of the criminal investigation.

In hunting down the source of this temporary aberration it was found that its time period 1952 to 1978 and its variation characteristics (start, stop, waxing, waning) are exactly the same as those of the air-borne nuclear weapons tests of the period, this being adequate to identify the

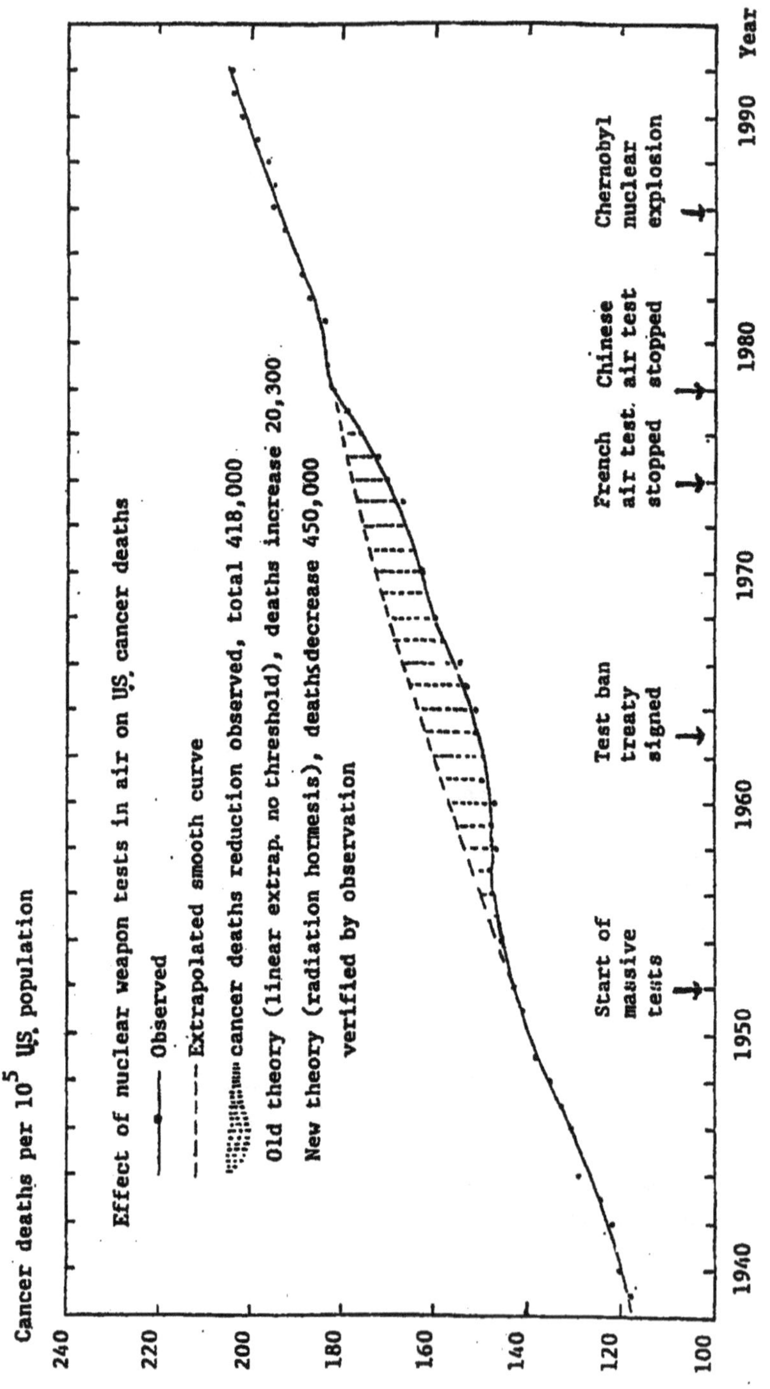
Cancer deaths per 10^5 US population
Effect of nuclear weapon tests in air on US cancer deaths
Observed
Extrapolated smooth curve
cancer deaths reduction observed, total 418,000
Old theory (linear extrap. no threshold), deaths increase 20,300
New theory (radiation hormesis), deaths decrease 450,000
verified by observation
Start of massive tests
Test ban treaty signed
French air test stopped
Chinese air test stopped
Chernobyl nuclear explosion
240
220
200
180
160
140
120
100
1940
1950
1960
1970
1980
1990
Year

Figure 2

weapons tests as the source of the aberration on a *prima facie* basis. After all there are very few things that can do anything about cancer.

The tests started in 1952, which is the starting point of the aberration of the curve. Nuclear radiation in atmosphere increased as test continued. Since radiation is known to cause cancer, an alarm was sent out and a Partial Test Ban Treaty was negotiated and signed in 1963. US, USSR and UK stopped air-borne tests that year but continued underground tests which the Treaty permitted. France continued tests in air, which stopped in 1974. Chinese test program stopped in 1978, which is the end point of the aberration period. The entire air-borne test period 1952-1978 coincided with the period of cancer anomaly and the test history is reflected in the ups and downs of the cancer curve. The tests thus provide us unwittingly a controlled experiment on cancer as related to radiation.

Falsity of Low Level Radiation Hazard

The great surprise is that what the test radiation did as shown by the teeth mark of the bite on the cookie in Figure 2 is to *reduce* the cancer deaths, instead of *increasing* them according to conventional wisdom (LEWT), which ironically brought about the test ban in the first place. Figure 2 shows the total number of lives saved from cancer deaths by the air-borne Nuclear weapons tests is about half a million (450,000 to be more accurate). Lucky serendipity! A calamity turned to a blessing! Thus it appears that radiation hormesis is indicated experimentally. More accurately, as far as the correlation of cancer with radiation is concerned, the experiment shows a negative correlation accurate to four places of significant figure. For the first time we have accurate data as the basis to draw important conclusions.

It should be noted that the number 450,000 is obtained from the interpolated dashed curve of Figure 2, representing the test free state. While such interpolation is routine in physics, the result following (450,000 lives saved) is so shocking that no one believed it and all tried to find excuses for explaining the teeth marks by esoteric reasons. A large amount of work was done to challenge a point that every physics freshman is doing routinely. This point will be returned to later.

Whereas positive correlation does not necessarily prove the truth of causation, negative correlation, as we have observed, does prove the falsity of the causation. Thus without further ado we have proved the falsity of increasing low level radiation causing cancer and the falsity of LEWT; and this applies not only to radiation from bomb explosions but also to radiation from nuclear plant explosions, nuclear waste dump explosions and so on, which include all nuclear hazard.

The Truth of Beneficial Effect of low Level Radiation

While we have debunked the falsity of the hazard of low level radiation, we have not precisely proved its beneficial effect. This missing point may be used to pull wool over the eye to confuse the issue.

The most common opposing argument is that something might happen in the period 1952-1978, such as improvement of health services, to reduce the cancer deaths, not necessarily the bomb test radiation. True, one graph such as Figure 2 is not sufficient. However, we are fortunate to live in a United States of 50 sovereign states, each of which can provide us a figure like Figure 2 here (breaking up the total cancer death rate into 50 rates for the 50 states). The 50 figures can tell much more than one. Six representative ones are shown in Figure 3. Here we have the display of evidence of criminal investigation of 50 cookie jars, each has a cookie with

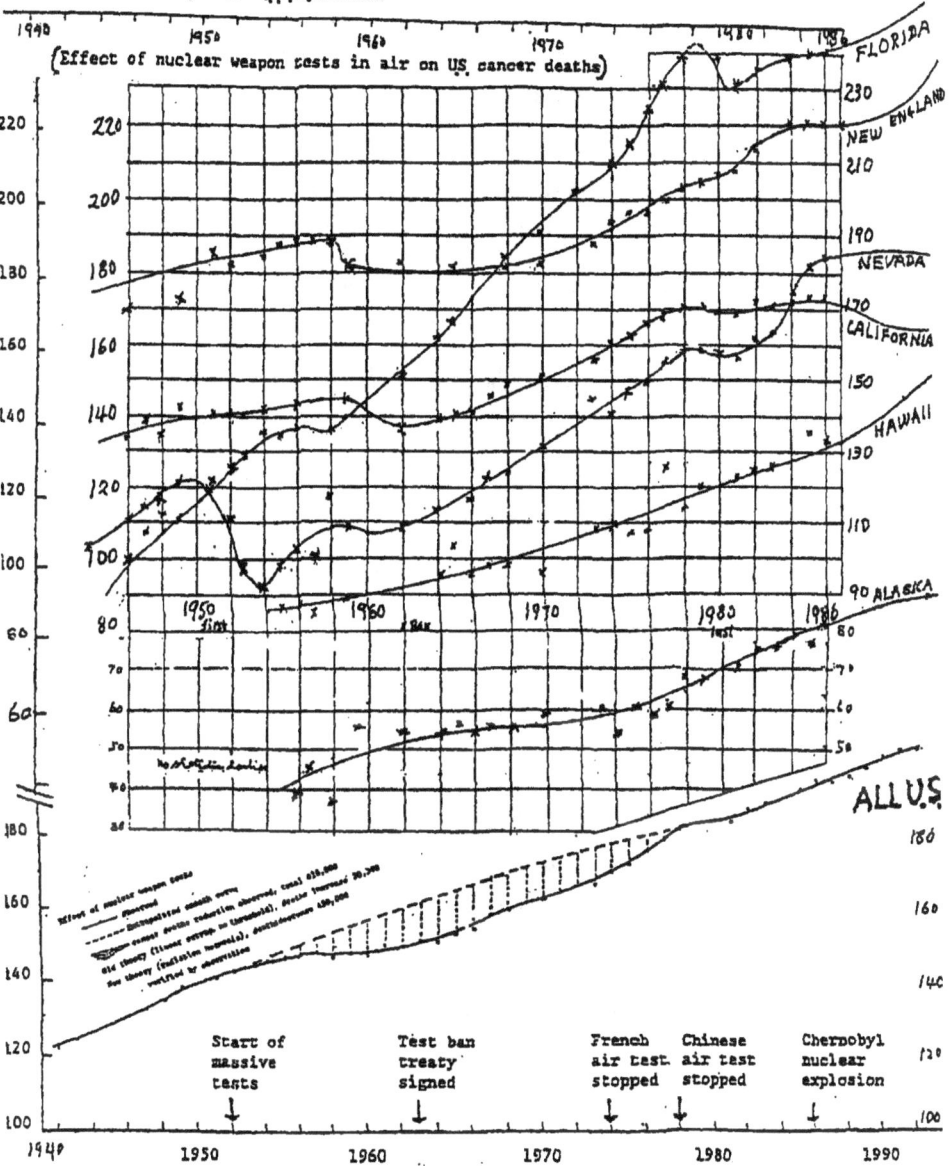

Figure 3

the characteristic teeth marks, telling us who, where and when the crime had committed.

Two curves, Hawaii and Alaska, represent the most distant and ocean separated states from the nuclear test site, and are expected to be the least affected by the tests. They should be close to the nuclear free "state of Eden" as represented by the dashed curve in Figure 2. They indeed are (see Figure 3). The use of the dashed curve as the nuclear free base line has been a serious bone of contention. The Hawaii and Alaska curves are a proof that the use of the dashed curve is justified.

Another two curves, Nevada and California, represent the states closest to the test site near Las Vegas, Nevada, which are expected to be affected the most by the tests. Indeed the Nevada cancer death rate was reduced by 25% from 1950 to 1954, during the beginning years of the test series, which is a medical accomplishment rivaling cancer hospitals. It is amazing that with all the interest in cancer, such a magic cure was never mentioned in any media.

One Editor responded to my submission of paper on this issue by saying that the nuclear tests brought in tens of thousands workers and changed the health statistics of the population of Nevada. Good Heavens! It takes only one person to push the button to test a bomb. The tens of thousands people were working in Oak Ridge, Tennessee, a thousand miles away, to *make* the bomb, not to *test* the bomb. Every high school student knows it. It is this kind of idiotic, ignorant and arrogant people without even a high school knowledge that controls our nuclear policy. God save America!

California was not so heavily affected because of mountains and prevailing wind direction. The California curve does have the tell-tale signature of the nuclear test teeth-bite marks.

Another group of two, Florida and New England States are in between the other two groups. The New England States show the same tell-tale teeth mark of nuclear test. Florida does not. The air over the Florida peninsula is largely from the ocean from both coasts. Notable is the period of contention 1952-1978 in which the Florida curve, presumably little affected by the test, is a simple straight line, indicating no esoteric happenings taken place during that period, which supports the use of the dashed line in Figure 2 as representing the test-free state.

These curves seem to indicate that an unknown source of cancer reduction has suddenly appeared in Nevada coincident with the nuclear bomb tests, which spreads out to other states and reduces the cancer death rate on the way with diminishing intensity as it propagates and then disappears at the time when the bomb tests stopped. This is the sufficient proof that the bomb test radioactive fallout is the agent of cancer reduction. The quantitative amount of cancer reduction in relation to the change of radioactivity in his study for the first time provides direct quantitative experimental measurement of the biological effect (cancer reduction) of low level radiation. The important conclusion is that low level radiation is beneficial.

This is the basic quantity to determine the safety level of all nuclear power facilities, which was not accurately known before. Only guess work was available (LEWT) and all nuclear safety standards were based on this guess work. LEWT implies all low level radiation is harmful and calls for extremely stringent safety measures. Our discovery here is just the opposite and the stringent safety measures are not necessary—as unnecessary as reinforcing all roofs of all houses to protect against a meteorite hit. The current nuclear regulations are completely outdated.

Radiation Hormesis

The reduction of cancer observed *may be* considered as the manifestation of radiation hormesis. However, as far as causation of radiation to hormesis is concerned we have only observed positive correlation which is not enough to establish causation without further ado.

To prove causation in general one has to do a controlled experiment in which the effect follows the cause immediately and is repeatable like a seismic explosion is proved to be the cause of seismic waves that follow the explosion. In our case the explosion of one test bomb may indeed be considered as a controlled experiment. However, the long term effects do not happen instantly and thus might involve other factors.

The only justification to consider a happening as a long term effect is that it is statistically repeatable in the long run. In our "experiment" the cancer reduction observed for each of the 26 years of the test period is the results of long term effects of all previous years. The average of the 26 years is a legitimate representation of the long term effect. A typical year, say, 1963, may represent the average since the fluctuation is small.

The observed cancer reduction of 1963, from Figure 2, is 7.3% of the 320,000 total cancer deaths of that year, which represents the effect of the bomb-induced-radiation of 30 mrem/year. This rate may be converted to *24.3% cancer reduction for an increase of radiation of 100 mrem/year (a doubling of background radiation)* for convenience of comparison. This is the quantitative law of the cause-effect relation of radiation and cancer mortality. This is the law of radiation hormesis we obtained experimentally which fills the void of our knowledge of low level radiation effects. The outstanding feature of the law is that the effect is beneficial and reasonably large.

This law of cancer reduction by radiation can be corroborated by other evidence. We know as a fact that Colorado has a radiation level double the background and its cancer mortality rate, shown by statistics, is smaller than the national average by 25%,[5] agreeing with our result 24.3% very well. Our result also agrees with Cohen's result of cancer reduction by increasing household radon level.[6]

From Figure 2 we can read out the total number of cancer deaths that were saved by the weapons tests, which is 418,000 over the 26 years. If there were no Test Ban Treaty, the 1500 bombs used in underground tests after 1963 would be exploded in the air and 1.2 million cancer deaths would have been saved, a great irony of the Test Ban Treaty.

According to the assumption of LEWT, the entire bomb testing period with an increase of radiation of 30 mrem/year would cause 21500 extra cancer deaths (0.3% of natural cancer deaths). This percentage is too small to detect experimentally, and will hardly show up in Figure 2. This may be the reason that no one actually compared the prediction of LEWT with the actually measured cancer mortality rates after the weapons tests, thinking that the effect is too small to be recognizable. And this turned out to be a great blunder—the effect is very large and in the opposite direction. It missed a chance of making a significant discovery and entailed tremendously large losses.

The Threshold of Radiation Hormesis

What is the threshold of the Hormetic effect—the radiation dose below which its effect becomes beneficial. To gain some insight on this important issue we may look at it from an evolutionary viewpoint. Our very existence on the earth is the proof that *the background radiation is hormetic*. If it were slightly harmful, the effect would be amplified exponentially in millions of generations in geological time to wipe out all lives on earth according to the laws of population biology.

The weapons test "experiment" was done at a radiation level of 30% above background. Any such small variation of radiation over the hormetic background can be expected to be hormetic and it actually is as we have shown above. The previously mentioned altitude effect of cancer rates shows the threshold is at least twice the background (the mountain states level), which is supported by voluminous cancer data. The Cohen study of household radon effect[6] suggests a threshold to be 10 times the background.

Beyond that cancer rate is no longer a sensitive gauge to study threshold. We turn to another gauge—the *longevity* as an indication of radiation hormesis (e.g. the atomic bomb survivors). The study of longevity will be published in a separate paper. Important information will be summarized here for the determination of the threshold.

The statistically most significant case on longevity is Kerala, India (population 29 million) which has a radiation level of 20 times the background. The life expectancy of Kerala residents is a remarkable 10.7 years longer than India (population 947 million) on the average.[7] Thus we conclude the threshold is at least 20 times the background. A similar case is Ramsar, Iran with a radiation level 120 times background. The observation of longevity in Ramsar indicates the threshold is at least 120 times the background (statistics unknown).

To show racial diversity does not affect radiation hormesis, the case of Law Chang, China (elevation 1800 m, doubled background radiation) may be cited. The life expectancy is 88 years, 20 years longer than Chinese average. Beyond humans, experiments on tissues indicate the threshold to be in the range of 100-200 times background.[8] Thus 20 times background is a safe minimum of the threshold, which is likely to be in the neighborhood of 100 times the background.

The threshold is thus at least 400 times higher and could be 2000 times higher than the safety level of radiation currently adopted—0.05 times the background for residents near a nuclear power plant. The current safety standard is unnecessarily high, which is one reason for the high cost of nuclear power. The high safety standard was introduced at a time of uncertainty on low level radiation to assure maximum safety. Knowing better now, nuclear power economy can be greatly improved.

Nuclear Accident and Waste Disposal

With the law of low level radiation established we proceed to study the major nuclear issues: the nuclear power plant explosion and nuclear waste disposal. The plant explosion can be studied readily. We have already shown that the weapons tests are a real life simulation of the nuclear plant explosion. Bomb explosion is the most efficient way to disperse radionucleides and its results are the maximum estimate of plant explosion risks. We need only take the bomb explosion results over with proper scale-down to fit our problem.

The consequences of a nuclear plant explosion had been studied extensively by the pro- (Rasmussen Report WASH 1400) and anti-nuclear forces with the investment of many man-years' work and publication of thousands of pages of reports still without conclusive answers to crucial points. The answers are now clearly given by the ultimate, weapons tests "experiments" with high accuracy.

The Chernobyl type nuclear explosion releases 10 times radiation of a nuclear bomb of the Hiroshima type. Today 414 nuclear power plants are in operation all over the world. If one tenth of them, say, 50 plants, were exploded together, the radiation generated would be the same as 500 bombs exploded, which was actually the amount in the weapons tests, the results of which are already known, that is, a reduction of 418,000 cancer deaths in the nation, which is not a risk but a benefit.

The Chernobyl consequences have been studied experimentally for 10 years and reported in the Conference *One Decade after Chernobyl*,[9] sponsored by the United Nations, the European Union and the World Health Organization. The conclusion of the conference is that there was *no increase of cancer deaths due to the Chernobyl accident*. The nuclear plant risk issue can thus be concluded.

A similar conclusion can be drawn on the nuclear waste issue. If one tenth of the world's power plants' nuclear wastes, from expend fuel rods to contaminated lab coats of nuclear workers, were "exploded" from the waste dump and scattered all over the world, the radiation generated cannot be greater than all the radiation from the explosion of an equivalent number of nuclear bombs, that is, 500 bombs, the results of which are already known in the present study. Thus no harm to the world will be entailed.

An accident involving one tenth of all the nuclear plants together is extremely rare but cannot be excluded, such as by the impact of a giant asteroid. Once such an ultimate accident is shown to be harmless, the nuclear power as a whole can be considered as safe. The protracted nuclear concern of 30 years may thus be ended.

References

1. T. D. Luckey, *Radiation Hormesis*, CRC Press, Boca Raton,1991.

2. B. L. Cohen, Int. Arch. Occup. Health, **66,** 71-75 (1964).

3. S. Kondo, *Health Effect of Low Level Radiation*, Kinki University Press, Osaka, Japan, 1993.

4. S. Gladstone and P. J. Dolan, *The Effects of Nuclear Weapons*, 3rd Ed., Dept. of Defense, Energy Research and Development Adm., 1977.

5. *Statistical Abstract of the United States*, Government Printing Office, Washington, D.C., 52 volumes from 1945 to 1996.

6. B. L. Cohen, Health Physics, **68,** 157-178 (1995).

7.United Nations, Department of Economic and Social Development, *Economic and Social Aspects of Population Aging in Kerala, India,* ST/ESA/SER/119, New York, 1992. [Microfiche 510. International Information Service (1993), #3080-M104].

8. Ludwig Feinendegan, Bull. Am. Phys. Soc. **42,** 1124 (1997).

9. United Nations, European Union, and World Health Organization, *A Decade After Chernobyl*, International Atomic Energy Agency, Vienna, 1996.

Long Life Expectancy as the Beneficial Effect of Low Level Nuclear Radiation

Peter Fong

Physics Department, Emory University

Atlanta, Georgia 30322

Abstract: Beneficial effects of low level nuclear radiation are found not only in the immune system against cancer, but also in the development system in extending life expectancy. Mortality statistics in the US indicate 7 out of 8 leading causes of natural deaths are reduced and life may be extended in the mountain states with doubled background radiation. The most outstanding example of increasing life expectancy is Kerala, India where background radiation is 20 times higher and life expectancy is 10.7 years longer. Exact proof is presented here to show that excess radiation is the cause of reduced cancer and other diseases and increased life expectancy. Nuclear safety standards are all outdated. Moreover, the "deadly" nuclear waste might turn into a gold mine; its radiation might simulate that of Kerala to prolong life, and of others to provide health bebefits. The most serious problem of nuclear power might thus be solved.

In the previous paper the same problem has been studies from nuclear radiation from air borne nuclear weapons tests. The radiation is the same but the space and time of its generation are under control and the results may be treated as a controlled experiment. This advantage is lost in this paper where we deal with natural background radiation which cannot be manually controlled. The problem thus becomes enmeshed in the issue of "ecological fallacy" which has bedeviled all studies of this kind. With

ingenuity and drastically new approach the problem is solved and the
results are exactly the same as before. Thus the two verify each other.

1. Introduction

It is well known that Kerala, India, has a high background radiation of 20 msv, twenty times higher than normal, due to the high thorium deposits. This radiation level is more than the long term radiation of 20 Chernobyl class nuclear explosion combined and we are interested in knowing how the people in Kerala respond to this severe environmental assault. According to one United Nations report,[1] life expectancy in Kerala is 10.7 years longer than in India and the trend has persisted for three decades for the time covered by the study. Before we rush to conclude the discovery of radiation as a longevity potion, we have to establish the causal relation of the two observations, lest the conclusion be challenged by such proposals as the high level of education or the three decades of local Communist rule as the origin of the longevity phenomenon.

In this paper we try to establish (1) a quantitative law that specifies the reduction of cancer mortality by increased background radiation, and (2) the extension of this beneficial effect on cancer to other diseases so that the total effect of the background radiation is the extension of life expectancy. Once this is done the Kerala result of life expectancy extension verifies our prediction on the beneficial effect of low level radiation and also establishes the threshold of radiation hazard to be at least 20 msv, or twenty times the normal background radiation. This threshold is so high that it renders most problems of nuclear power safety and nuclear waste disposal trivial.

2. Background Radiation and Cancer Mortality

The first observation of cancer reduction by increased background radiation was made by Willard F. Libby, who, as an Atomic Energy Commissioner, spoke to the American Physical Society[2] that cancer rate at Denver, which has doubled background radiation, is actually lower than in Los Angeles, thus discounting the risk of low level radiation. Though it indicated a beneficial effect of low level radiation, the idea was so outlandish at the time that it was never pursued in earnest, a costly mistake in retrospect. Nevertheless it did establish the background radiation of 100 mrem per year as the guide post of nuclear radiation safety. Incidentally the fact only relates cancer to altitude. To identify the altitude effect with background radiation and with no other factors is a problem we will pursue.

With forty years additional statistical data[3] after Libby, we can easily generalize his conclusion from Denver to the entire Rocky Mountain region. The statistics establish that the cancer mortality rates of the 8 mountain states (Arizona, Nevada, Idaho, Montana, Utah, New Mexico, Colorado, Wyoming) as a whole are lower than the United States average by 25%. The reduction ratio 0.752 is nearly a constant over the past 40 years with a root-mean-square deviation of 2.5%. These results are displayed graphically in Figure 1. The beneficial effect of low level radiation (radiation hormesis) has been studied widely and has been reviewed by Cohen.[4]

It is natural to try to establish a quantitative law on the cancer reduction by low level radiation. Many attempts have been made but all are severely criticized by the epidemiologists and none is accepted. The

ST/ESA/SER.R/118

Department of Economic and Social Development

Economic and Social Aspects of Population Ageing in Kerala, India

United Nations   New York, 1992

- 13 -

Table 8. Expectation of life at birth: Kerala and India, 1951-1981

| Period | Kerala | | | India | | | Differen[ce] |
	Both sexes	Males	Females	Both sexes	Males	Females	Both sexes
1951-1961	44.8	44.3	45.3	36.0	36.4	35.7	8.8
1961-1971	55.7	54.1	57.4	43.8	44.1	43.6	11.9
1971-1981	61.6	60.6	62.6	50.9	51.6	50.2	10.7

argument is that cancer is determined by many factors—radiation is one and there are many others such as smoking habit. The observed cancer mortality rate M is a function of all the relevant factors $M(x_1, x_2,...)$. The contributions of the separate factors $x_1, x_2, ...$ to M have not been separated and therefore a function $M(x_1)$ of a single variable x_1 such as radiation level cannot be established. It is argued that a slight change of the smoking habit could mask the effect of radiation and therefore from the observed M one cannot establish $M(x_1)$, which is the desired result.

This epidemiologist's objection, with their 100 confounding factors, has thwarted all efforts to establish a single variable correlation, which is needed in setting up regulation rules and safety standards. As a result safety standards are set up by the principle of maximum possible safety which is notoriously wasteful and thus is unnecessary. The basic problem must be solved first to avoid ridiculous waste.

The problem of separating the contributions of different variables occur often in theoretical physics. A typical solution is to separate the contributions by their orders of magnitude. It so happens in nature that contributions from different factors vary widely in orders of magnitude so that they can be separate without ambiguity. For example, the electronic, vibration and rotational energies of a molecule differ by two orders of magnitude each and thus can be separated without problem. If the orders of magnitude are comparable, this method cannot be used. But the probability of two phenomena having the same order of magnitude is extremely small in nature.

Granting the orders of magnitude are different, the problem of separating $M(x_1,x_2,...)$ into $M(x_1)+M(x_2)+...$can be done by the method of successive approximations. Solve the first order problem first. Have the results subtracted from M and the remainder is of the second order. Then solve the second order problem. And so on until all orders are solved, which represent the complete solution of the problem.

For example, Newton solved the first order problem of the elliptical orbits of the planets using the largest source of gravity force of the sun. One hundred years later Laplace solved the second order problem of the remaining minuscule effect of the advance of perihelion of the planetary orbits by the much weaker gravity force of the planets themselves. Finally another 100 years later Einstein solved the third order problem of the last remaining infinitesimal deviation of the Mercury orbit by the weakest effect of general relativity. Together we have a complete and all-inclusive solution of the planetary problem. If we should start with the Mercury orbit in its minute details, we cannot solve any problem and we will remain in the middle ages without the scientific revolution.

Cancer mortality rates depend on many factors. Radiation can change it by 25%. Smoking habit can change it by 16% (smokers have a life expectancy 16% shorter). Pollutants in the environment can change it by 1%. These are all the major factors we know that influence cancer rates. They do not vary as widely as the gravity forces of the sun, of the planets and of the general relativity, but they do vary. With judicious treatment of the empirical data there is hope to separate the variables. A condition to our favor is that cancer is a very stubborn disease not easily amenable to change. That is why it is so difficult to cure. Thus there is no wild card that may upset our planned study.

To follow the procedure of successive approximations we must start with a firmly established zeroth order approximation. Then deviations from it may be attributed to higher orders of approximations. For this we make use of the firmly established fact in Figure 1 that the 8 mountain states of the US as one body have a cancer rate lower than the national average by 25%. The independent variable is the altitude.

This is a zeroth order law with very crude result. We now proceed to the first approximation to make it more precise. We hypothesize a linear law of cancer reduction according to the altitude and try to establish its mathematical form by the empirical data. This is the most natural step to take once the zeroth order result is established. To solve the first order problem, we treat the 8 mountain states separately (instead of as a block in the zeroth order approximation) and try to fit their cancer rates with their altitudes by a linear law. The 8 states have altitudes varying from 2000 ft to 6000 ft which are used as the abscissa. The cancer mortality rates of the 8 states are plotted as ordinates (see Figure 2). If the data points fall on a smooth curve then the hypothesis is verified and a law on cancer rate versus altitude may be established. If there are deviations from the smooth curve we may treat them as the remainder of the first approximation and try to explain them by the second order studies in the next approximation.

Preliminary study (can be seen from Figure 2) shows that the points generally follow a declining smooth curve, supporting our hypothesis. But there is a marked deviation in the case of Utah, where the data point is way below the smooth curve, indicating a drastic reduction of cancer rates by 1/3. This deviation can be easily explained by the fact that there is a large population of Mormons in Utah whose non-smoking and non-drinking life-style certainly help reduce the cancer rate. In this way the deviation can be explained by the life style factor in the second approximation we anticipated earlier.

Cancer Ratio: Mountain States/ Entire United States

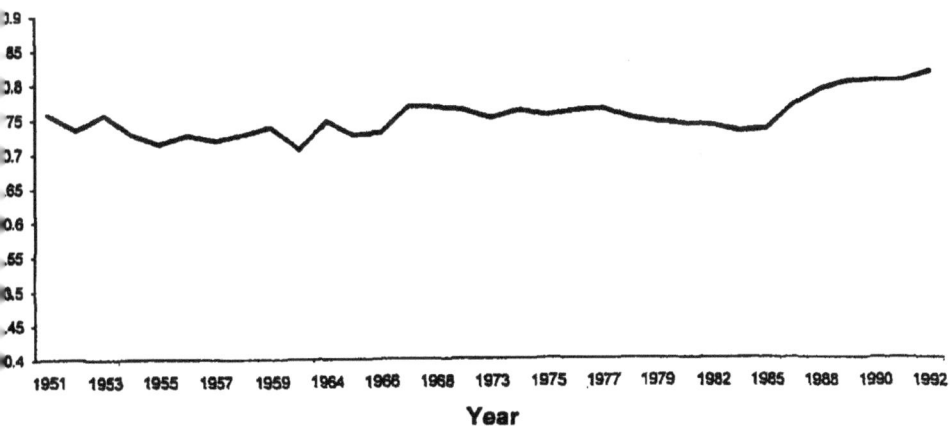

Figure1

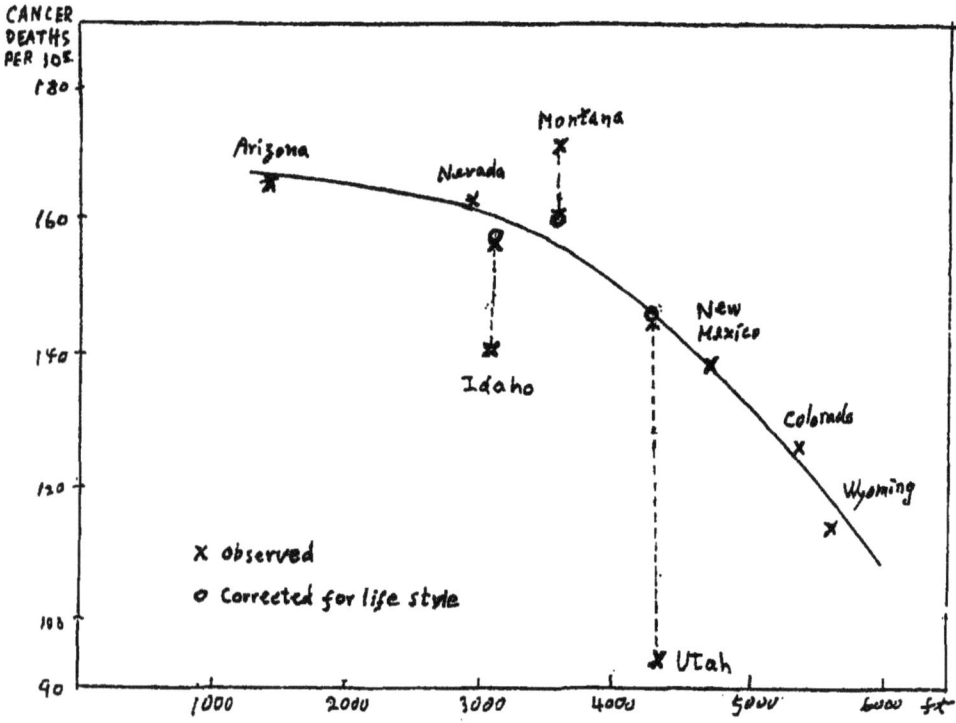

Figure 2

Therefore our approach is in the right direction. Moreover, it accomplished the expected objective of separating the variables: altitude in the first approximation and life style in the second approximation so that a law of a single variable (altitude) may be obtained. (In principle the method may be extended to all orders to find all laws of single variables but in practice other expedient approaches may be more effective.) To pursue this approach the detailed treatment should be improved to make the first approximation as accurate as possible.

One obvious improvement is the treatment of the altitude. The altitude of a state should not be the geographical average altitude of the state; it should be the population weighted average of the altitudes of all residential municipalities of the state. The populations and altitudes of all the municipalities down to the village level of all the mountain states are available from published sources.[5] The calculation of the weighted average altitude is straightforward but tedious. It is necessary to improve the accuracy of the first approximation. The calculation was done by Mr. David Bovi, whose help is kindly acknowledged. The results are used in the presentation of Figure 2, which are better than the preliminary results not shown here.

The results displayed in Figure 2 show good correlation of cancer mortality rates with the average altitudes of the 8 mountain states so calculated. The most outstanding exception is still Utah which has been noted and explained earlier. Now we have to make quantitative correction of the data point for the Mormon life-style effect. The only quantitative value we know of the reduction of cancer mortality in this respect is that of abstinence of smoking, which is 23% from independent smoking studies. Actually Utah, as seen in Figure 2, has a reduced cancer mortality rate of

35%. There is an extra 12% due to other factors than no-smoking, which include no drinking and other austere and devout practices together with whatever cooperative effects of all of them working together. Since we have no independent quantitative determination of these contributions, we choose to assume them to be 12% as observed and try to justify the assumption indirectly later.

We next consider the deviation of Idaho. Idaho has a minor population of Mormons which caused the smaller deviation from the smooth curve. The Mormons of Idaho are the largest of 6 religious groups of the State. We therefore estimate the Mormon population to be 1/3 of the State. The deviation of Idaho's data point is thus 1/3 of that of Utah. The corrected data point is plotted in Figure 2 which is very close to the smooth curve.

The only other deviation of the data points from the smooth curve is Montana, which is much higher than the smooth curve, indicating higher cancer rate. Since Montana is two-thirds in the plain and one-third in the Mountain, it was first thought that the rule of the plain states would prevail (high cancer rate). However, a careful examination of the population weighted average altitudes of Montana municipalities shows it to be 3603 ft and the mountain state rule would apply (low cancer rates). Therefore other explanation must be sought for the abnormally high cancer rate of Montana.

A check of Montana's mortality rates of other major diseases[3] shows that they are either the highest (such as cancer and stroke) or the next highest (such as heart disease) among the 8 mountain states, just the opposite of Utah which has the lowest rates in just about all categories of diseases. The suspicion is thus that Montanans pursue an indulgent life style opposite to the Mormons of Utah with higher than normal rates of consumption of tobacco and alcohol. It is natural to seek answer from life style in the second approximation.

But there is no direct statistics available in this regard to check the speculation. It would take a long time with a large amount of manpower to canvass the Montana population for their smoking and drinking habits; and the result obtained can hardly be accurate to 50%. A similar problem occurs in a CDC (Center of Disease Control) project to study the same problem. Although the preliminary results were promising (they have shown the existence of the *dip*), they did not pass internal review of CDC for lacking an account of the smoking effect and the project was abandoned. The fact is that no one knows the smoking effect and how to give an account in the research program. Difficult problems of this kind often occurs in practical issues (not in home work exercises) and they are the test of the mettle of the empire builders.

In this case a perceptive observer can recognize that a piece of relevant information is readily available in the tobacco and alcohol tax collected, which is perhaps the best proxy measurement of the degree of smoking and drinking, far more accurate than canvassing and the numerical result is published by the government routinely without knowing its serendipitous value. The taxes collected for tobacco and alcohol[6] in Montana is $33.7 per person per year. This is the second highest among the 8 mountain states (next to Nevada, which lives by tourism) and is 23% higher than the average of the 8 mountain states, $27.4. Thus it is quite reasonable to say that the excess 23% of smoking and drinking caused the abnormally high cancer rate and the deviation of the data point above the smooth curve.

The amount of deviation of the Montana point in Figure 2 can even be proved quantitatively—it should be 23% of the deviation of the Utah point and Figure 2 shows that it is actually 29% of that of Utah, the agreement being reasonable enough. This explanation is desirable because we attribute all deviations of the first order effect (radiation) to the known second order effect (life style), thus unraveling the Gordian knot of the effects of confounding causes by successive approximations.

That the Idaho and Montana points are close to the smooth curve justifies our assumption of 12% for the undetermined life style effects. If the assumption of 12% is off by a large factor, both of the Idaho and Montana points will be far from the smooth curve. With all deviations from the smooth curve removed, the problem is well solved in the second approximation.

The smooth curve of Figure 2 is nearly an exponential curve correlating the cancer mortality rate with the average altitude. Since background radiation decreases exponentially when altitude decreases (see Figure 2 upside down), the correlation of cancer mortality rate with radiation is thus very closely a linear proportionality, which is desirable. On this basis we establish the second approximation of the quantitative law concerning cancer reduction by low level radiation, that is, *a 26% reduction of cancer rate under the exposure of a doubled background radiation in a linear correlation.* The result (accurate to 1%) is not much different from that of the first approximation (accurate to 10%) of a reduction of 25% but the agreement is fortuitous because two large corrections happen to cancel out each other.

The effect of the linear law is profound. Strictly speaking the only thing empirically established is that the factor that determines the mortality rate is one that varies exponentially with altitude. The altitude is just a proxy of the true cause. Any supposed cause that does not vary with the proxy can be excluded immediately. This includes most features of high elevation such as fresh air, weather and climate, which do not vary in proportion to elevation.

However, any factor that varies in proportion with the elevation cannot be excluded as a possible cause. For example, the atmospheric pressure and its consequent boiling point of water and so on do vary in proportion to elevation. Of special importance is the point that there is an oxygen deficit of about 16% at high altitude, which may represent 16% less oxygen free radicals and an equal amount of increase of immune capability. This capability increases with elevation and therefore *could* be a cause of cancer reduction in proportion to elevation. How to eliminate this possibility?

While this effect increases in proportion to elevation, it is not exactly linear. Thus a fine point on linearity makes it possible to distinguish the two cases of radiation and atmospheric pressure. The atmospheric pressure *decreases* exponentially when elevation increases but the radiation *increases* exponentially under the same condition. In a rough approximation both can be approximately as linear and thus indistinguishable. But in exact presentation the difference will show up. If the cause is radiation the curve in Figure 2 should be an exponential curve concave downward. If the cause is atmospheric pressure the curve should be an exponential curve concave upward. The curve we obtained in Figure 2 is indeed concave downward and this proves that the cause is radiation and not atmospheric pressure or its cohorts. Thus the statistical study makes it possible to pin point one single cause for the cancer reduction, that is, the variation of background radiation exclusive of any other causes. This is an exact proof and no further follow up is necessary. Aside a direct experiment (such as in the previous paper), this is the only proved truth of the low level radiation health effect. Again the fine discrimination of concave upward and downward hinges on the use of the population weighted average altitudes and will not be certain in other statistical studies.

This success is based on the fact that we use the largest possible, the ultimate, data set including the entire population, living and dead, in the United States of 250 million in 50 states covering a continent spanning over half a century. Any truth cannot escape this all inclusive data set and can be sifted out with proper mathematical procedures.

Furthermore, the truth so obtained is the absolute truth which cannot be challenged by any further experiment because any further experiment can only be based on a sub-set of the data, which cannot contradict and override the truth of the entire set. No further study and research are needed. Any theory, practice, popular opinion, and so on, past, present and future, that are contradicted by the law obtained here must be abandoned. This includes the theory of linear extrapolation without threshold and the entire edifice of nuclear safety standards and regulatory procedures adopted at the present time.

It is important to note the significance of the zeroth order approximation. Without it we have no mandate to hypothesize a linear law and to fit it with empirical data; the whole scheme would be a castle on sand. The zeroth approximation is based on the existence of a Rocky Mountain Range which stands out sharply on an otherwise flat topography, thus establishing an indisputable zeroth order approximation to begin a study of successive approximations. Without it the Gordian knot has no lead to begin unraveled. The pessimism of the epidemiologists would prevail. The spell is broken because of the existence of the Rocky Mountains, which is an accident of creation, not included in the design of the epidemiologists.

The mortality statistics definitely establishes that the increase of background radiation is the cause of the cancer mortality reduction and all other possible causes are eliminated. The endless discussions on all issues of ecological fallacy can now be put aside. The linear law so established, which implies a threshold of low level radiation hazard of at least a doubling of the background radiation (100 mrem a year), can then be used to deal with other practical applications, such as nuclear power plant accidents and nuclear waste disposal. The safety standard so set up will be practical and economical in the sense that we sacrifice exactly what must be sacrificed, not a blank check on an upper limit with countless unnecessary sacrifices. In other words the law is black and white, not gray. The latter was done in the past because of ignorance. It is unnecessary and unaffordable, and can be corrected now.

In Figure 3 the exponential curve for the natural background radiation plotted in the ordinate scaled on the right against the elevation as abscissa is compared with the cancer death rates of the mountain states scaled on the left. The death rates corrected for life style fit the exponential curve well. This may be considered as the quantitative proof of the reduction of cancer deaths by background radiation. The result agrees with that of a similar study reported in the other paper based on radiation from nuclear weapons airborne tests. With all the quantitative results in agreement, this major problem on nuclear safety is now solved

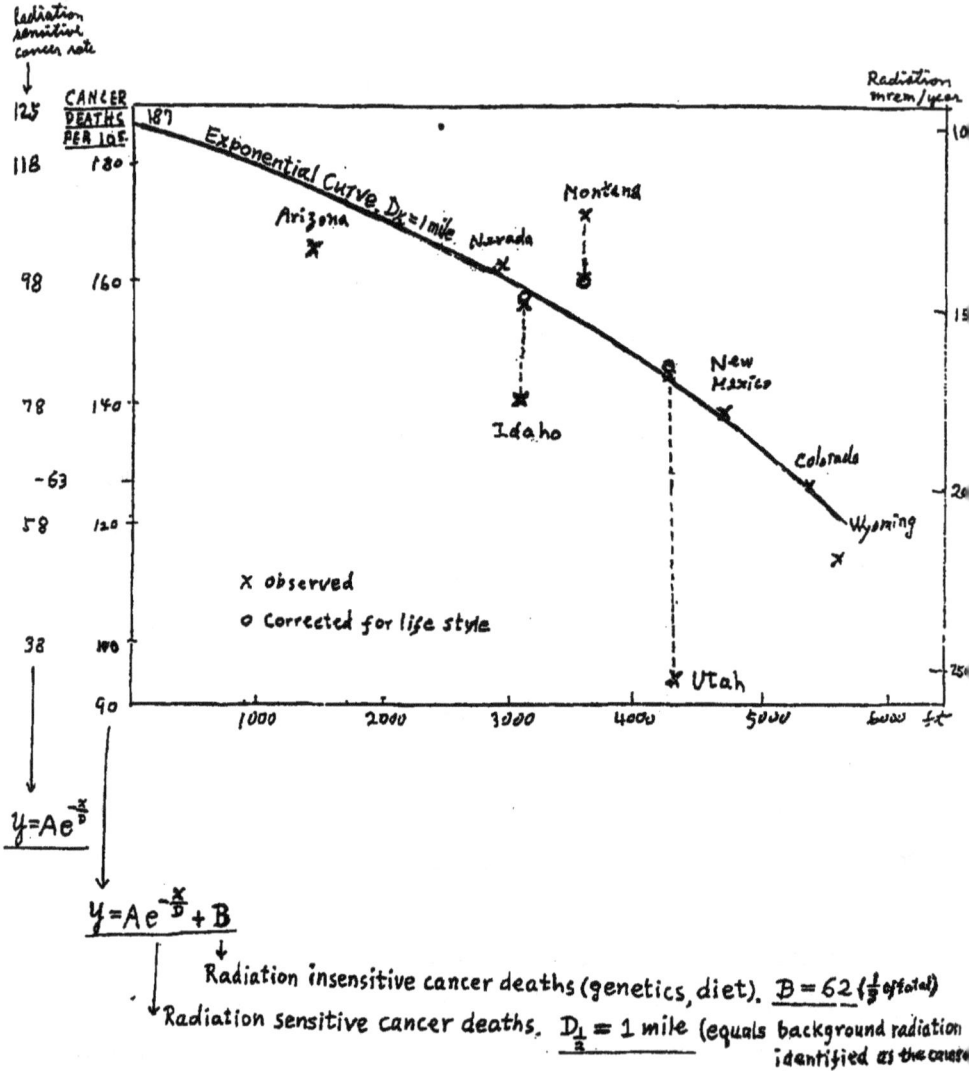

Figure 3

beyond reasonable doubt and cannot be overturn because they are based on the complete health statistics of the entire United States.

With the guide of the exponential curve, the cancer death rate given in Figure 3 may be represented by a simple formula given in Figure 3 which shows an exponential term as expected but also an added constant term B which represents cancer deaths not related to altitude (radiation). This should include cancers due to genetics and diet, which we know are possible sources of cancer but have no idea of its magnitude. From the value of B we conclude on the sea level (for most of the population) one-third of the cancer deaths belong to these categories. This is a happy windfall. A serendipitous harvest. It is not by sheer good luck. A complete solution of a problem ought to contain most important conclusions.

It would be remiss not to take advantage of this discovery to touch on several points of general interest. It has been speculated that the meat based diet is a source of cancer. Study of fresh immigrants from China with a rice based diet show much smaller cancer rates than Americans. However, old timer immigrants accustomed to American diet show cancer rate comparable to Americans. What's wrong with meat based diet? It is rich in hormones. Female sex hormone promotes breast cancer. Male sex hormone promotes prostate cancer. The two are leading causes of death for the respective sexes. With all the interest on cancer and our conclusion that up to one-third of the cancers could be of this origin, it is a problem that deserve better attention.

A piece of related work on low level radiation is the study by Cohen[7] on the effect of household radon which encompasses dozens of confounding factors but is still debated by epidemiologists. He found that increased radon level is correlated with decreased lung cancer mortality, which is still challenged by the epidemiologists. From our point of view our established law should apply, leading to the same trend up to a radon level that is twice the background radiation, all epidemiologists' challenge to no avail. After conversion of units and changing the lung cancer mortality rate to all cancer rate, Cohen's numerical result of the proportionality constant agrees with ours, thus his result is verified by our law. Cohen's linearity result extends to 10 times the background radiation. We may thus argue that the threshold of radiation hazard is at least 10 times the background radiation.

We now come to the Kerala phenomenon with a radiation level 20 times the background radiation. From the longevity point of view of Kerala, 20 times background radiation cannot possibly cause any health harm, otherwise people cannot live 11 years longer. Thus the Kerala fact indicates that the radiation hazard threshold is much higher than 20 times background.

Twenty times background is still a lower limit of the threshold. How high the threshold will go? Based on the study of radiation effect on tissues, Feinendegen[8] put it approximately at 100 to 200 times the background. Based on the Kerala facts on humans, twenty times background seems fail-safe from a practical point of view. This threshold is 400 times higher than currently adopted for resident living near a nuclear power plant. Such unnecessarily stringent standard wasted hundreds of billion dollar.

3. From Cancer Reduction To Longevity

The Kerala phenomenon shifts the attention from cancer reduction to increase of life expectancy. The former is a necessary condition but not the sufficient condition of the latter. It is interesting to explore the connection between the two.

Longevity requires the reduction of cancer and all the degenerative and communicative diseases. The mortality statistics of the US listed the breakdown of mortality into 10 leading causes of death. Two of them, accidents and suicide, are artificial, and will not be concerned here. The other 8 include, besides cancer, 6 degenerative and 1 communicative diseases. If the mortality rates of all of them are reduced then by definition longevity is assured.

While studying the mortality rates of cancer of the mountain states, a casual look over the mortality rates of the mountain states of other leading diseases causing death reveals that they are mostly also lower, indicating a significant connection among the rates. The low rates of all diseases imply a longer life expectancy for the mountain states which is an interesting result. However, this natural long life of the mountain states is compromised by accidents and suicides and did not show up in vital statistics.

The Kerala phenomenon makes it important to consider life expectancy as a scientific problem. As a result a study similar to that leading to Figure 1 for cancer is carried out for each of the 8 leading diseases causing death which includes heart diseases, cancer, stroke, degenerative diseases of the lung, pneumonia/flu, diabetes, liver diseases, and artery disease, based on published statistics.[3] The results in graphical form are presented in the following paper. Results of separate diseases are generally the same as those of cancer. The most important conclusion is that nearly all mortality rates of the 8 diseases are reduced in the mountain states (thus by radiation) with comparable proportions as that of cancer. Reduction of all leading diseases causing death means longevity. This establishes the connection of radiation with longevity as displayed in Kerala.

Nevertheless there is one, and only one, outstanding exception of the 8 diseases—the degenerative diseases of the lung excluding cancer and communicative diseases of the lung. We have already mentioned previously that there is an oxygen deficit of about 16% in the mountain states. Therefore the lung must have over worked to make up the deficit. The oxygen free radical is thus not decreased and yet the wear and tear will increase the degenerative disease of the lung. The exception actually proves our point.

Having the attention focused on life expectancy, we collected information on longevity for further study. Kerala is the most outstanding case. Statistics show that life expectancy of Kerala in 1951-1961, 1961-1971 and 1971-1981 are 44.8, 55.7, and 61.6 years respectively, whereas in India as a whole they are 36.0. 43.8 and 50.9 years respectively, representing an increase of 8.8, 11.9, and 10.7 years respectively for the three decades for Kerala over India.[1]

There are corroborative evidence verifying the conclusion of radiation effect on longevity, not just the well-known effects of reduction of cancer. The radiation effect of the atomic bombs exploded in Japan has been thoroughly studied.[9] The curve of mortality rate (of all deaths, not just cancer) as a function of lifelong radiation exposure does not have a minimum at the origin as expected from the theory of linear extrapolation without threshold but at a substantial dosage several times the background radiation, indicating a beneficial effect of radiation on longevity (not merely cancer reduction). In other words, they have shown the *dip* of the curve. This happened to residents living at points far from ground zero (>5000 m) of the atomic explosions. Near ground zero the mortality rates are naturally greater because the high level radiation caused lung cancer and leukemia.

Another piece of information to verify the same conclusion from an entirely different source is the study of the radioactive luminous watch dial painters. It was well known that many of such painters died of cancer from the radioactive material in the dial paint. But it was also found that there is a small group of painters with low degree of exposure who lived to ripe old ages.

A third is a recent report[10] that there is a mountain village (elevation 1800 m) Law Chang in Kweichow, China where the life expectancy of the one hundred odd families is 88 years. This verifies the conclusion of longevity expectation at high altitude of the American mortality studies if the accident (largely due to automobile) and suicide death rates are negligible as in Law Chang. This unusual fact has attracted scientific attention and investigation. The Geochemistry Research Institute of the Academy of Science of China in Beijing has analyzed the chemical composition of the water supply and found it low in mineral and low in sodium. No unusual chemical was found which might act as a longevity potion. No one suspected of the effect of high background radiation (doubling the sea level value at 1800 m altitude, comparable to that of Denver, Colorado). In all the above three cases the subject matter of study is longevity which occurs on the concurrence of many independent causes which are not ordinarily expected.

Having established a qualitative correlation of radiation and life expectancy, we try to gain some understanding of the quantitative connection of the facts at Kerala that 20 fold increase of radiation is correlated with 10.7 years increase of life expectancy without elaborate, exact mathematical calculation, which may not be possible in the first place. If the law established in Sec. 2—a doubling of radiation level reduces the cancer rate 26%—is extended to higher radiation levels, cancer would be nearly completely eradicated before radiation reach the Kerala level of 20 times the background. In the same way statistics (see the next paper) indicate the same for all other diseases. Taking it at the face value we combine it with the commonly stated conclusion that a complete eradication of cancer only increases life expectancy 2.5 years. Since cancer accounts for 24% of all deaths, the increase of life expectancy at the Kerela radiation level would be 2.5/0.24 which equals 10.4 years. This is not a proof of the actual value of 10.7 years but will lend some credence to it.

4. Theoretical Explanation

We now consider the cause of the increase of life expectancy. The hormetic effect of cancer reduction may be understood as a sort of immunological effect such as a polio shot of weakened polio virus may stimulate the immune system against polio. Cancer is due to DNA damage and malfunction, which occur naturally. Since life began, single cell organisms learned to repair DNA damage, otherwise they would not have survived to this day. A simple negative feedback mechanism, which occurs naturally in the evolution process, may cause an assault on DNA to react to stimulate the repair mechanism to heal the damage.

However, immunity to bacterial and virus diseases are specific—polio vaccine just for polio only. The radiation induced immune reaction in this way may explain the reduction of cancer caused by radiation damage of DNA, but how can the radiation help ameliorate degenerative diseases which are not specific and are not related to bacteria and virus, and are a natural physiological process?

One view is that there may exist a general immune reaction to produce a chemical reducing agent to scavenge the oxygen free radicals that cause diseases. Ionizing radiation passing through tissues may produce ionization resulting in the production of oxygen free radicals which are so reactive that they may produce DNA damage. On the other hand degenerative diseases are usually attributed to the declined ability of the aging body to remove the oxygen free radicals (thus using Vitamin E as anti-oxidant to reduce aging) so that the functions become degenerated. According to this view the radiation assault stimulates this particular immune reaction to generate scavengers to sweep off oxygen free radicals. The result is (1) reduction of cancer (actually preventing it from happening) and (2) amelioration of degenerative diseases to slow down aging and thus to prolong life expectancy. According to Pollycove[12] trillions of mutations in our body are controlled each day by such a highly efficient DNA damage-control system, only a billion are induced by background radiation, the majority of mutations being generated in the natural metabolic processes.

This is not the only possibility of radiation induced longevity. Another view is that the radiation assault changes not only the immune system but also the developmental system that regulates the progress of the life cycle. One example is the fact that women in mile-high Denver have their menopause delayed by one year. This may represent the stretching out of the life cycle by one year and this is not related to the immune system. Since Denver has a doubled background radiation, Kerala, with 20 times background radiation, may thus have the life cycle stretch out by 10 years and thus an increase of life expectancy of 10 years.

Another more practical evidence is the fact that China has produced and marketed prawns that grow 15% to 20% larger under the stimulation of low level radiation. This is strictly a growth problem, not related to the immune system. The possibility is that the radiation may interfere the growth system by increase the release of the growth hormone. Some amount of basic research on this phenomenon has been carried out in China[13] but much remains to be done. It seems that low level radiation not only affects the genetic and immune systems but also the growth and developmental systems, which are important to embryology and gerontology.

Besides the growth hormone, the pineal hormone may also be involved. The latter (melatonin) is released during the REM (rapid eye movement) sleep stage which is severely depleted in the old age (gerontological insomnia), and thus the shortage of melatonin has been related to aging. The pineal body has appeared in the fish stage of phylogeny development and has not changed much since. Melatonin is the only hormone it produces. It seems to control the overall growth process including the timing of puberty and mating (and presumably menopause and aging). This hypothesis explains the overgrowth of the prawns and the continued growth of an adult to delay menopause and to staff off aging.

5. Experimental Tests

This hypothesis can be easily checked by experiment with laboratory mice. Expose them to low level radiation from high altitude, or from the uranium mine tunnel, or from artificial sources such as reactor neutrons, and find out if they grow bigger and live longer than the control group. If they do, then study their brain hormone (pituitary and pineal) release rate and compare with the control group. Also study the change of the immune system. If results are positive, it would have tremendous impact on medicine, gerontology, and human life.

The hormetic effect of radiation on cancer is likely to be duplicated in carcinogens. Both deal with the development of cancer, the only difference being the external agents that bring it about—one is physical and the other is chemical. The mechanisms of the processes could be similar and the outcome could be comparable. Indeed hormetic effect of low dose dioxin (a noted carcinogen) on reduction of most forms of cancer on mice have been reported.[14] Studies of chemical stimulants on longevity are desired. They may eventually lead to drug treatment of aging. Similar studies on chemical toxins and environmental pollutants may ascertain the threshold of their harmful effects to determine effective and reasonable safety standards.

Since longevity has no survival value to the species (the old is expendable), the natural evolution process does not take advantage of the golden opportunities, if any, to extend life span, which are wasted in the history of evolution. Human efforts to take advantage of these opportunities, such as we have discussed here, may be extremely fruitful and easily achievable, compared with such efforts as improvement of vision, on which evolution has done the job exhaustively and not much further advancement can be expected.

6. Afterthoughts on Nuclear Energy

The strikingly large beneficial health effect in Kerala outdates all safety standards of nuclear power operations. Indeed the Chernobyl nuclear reactor explosion has been found not to lead to the predicted increase of cancer deaths.[11] The nuclear power accident risks have been greatly over-stated.

The radiation released in the Chernobyl meltdown and explosion is 10 times of that of the Hiroshima atomic bomb. In all the atomic bomb tests in the atmosphere involving 400 bombs before 1963 when the Test Ban Treaty was signed, plus an additional 100 air bombs tests up to 1975 when the test fever stopped and plus further an equivalent of 100 bombs from the 1500 underground bombs tested from 1963 to 1990, the total 600 bombs exploded in the air generated, according to experimental measurement, 30% increase of the background radiation in the air (30 mrem/year) during the peak test period, which has declined to zero at the present time.

The 600 bombs are equivalent to 60 Chernobyl type explosions. If all the 414 commercial nuclear power plant of the world today were exploded like Chernobyl at the same time, the total increase of radiation would be 2.07 times the background radiation at the peak period. The Kerala radiation level is 10 times greater than that with a beneficial effect of 10 years additional life. One tenth of that radiation would increase life expectancy of one year. Thus one year's additional life is the result of the nuclear cataclysm of all reactors meltdown at the same time. The much advertised nuclear disaster is not so much an apocalypse but rather a millennium. Since longevity means the absence of all health hazards, this is the ultimate proof that nuclear power is completely harmless under all possible extreme conditions (except the short term effect, which is common to and not different with all explosions). Furthermore the erstwhile most difficult nuclear issue—the waste disposal—now appears not only trivial but also potentially beneficial. Trillions of dollars may be saved.

From the air-borne bomb test radiation data, we can calculate the radiation level increase of one Chernobyl accident to be 0.5% of the background radiation. Even we use the old theory of linear extrapolation without threshold we can calculate the cancer death rate of Chernobyl to be 5 deaths a year. But the previous nuclear risk calculations (pro and con, ca. 1975) of cancer deaths of one nuclear meltdown are always hundreds or thousands a year. Those results are now known to be factually wrong. However, the mistake is not the choice whether low level radiation is harmful or beneficial, but a gross over-estimate of the amount of radiation generated, equal to about one time of background radiation instead of 0.5%.

One source of this over-estimate is the radiation of the short lived nuclides during the first few days of the accident. This part, died out quickly, does not appear in the bomb test radiation amount. But the damage of this part can be avoided by evacuation or staying in shelters for a few days so that it should not be counted. A second part is the over-estimate of the source-term discovered by the American Nuclear Society (1984) and the American Physical Society (1985) for the amount of radionuclides ejected out of the accident site. A third part concerns the disperse of the radionuclides in the environment. It is the use of the troposphere distribution which applies to small accident with small release, whereas in large accident with large release such as in Chernobyl, the radionuclides distribute through stratosphere. In this case only the long-lived nuclides survive so that the radiation level is much reduced.

Scavenging by rain and wind in troposphere is very effective. In urban areas most nuclides are swept to the hydrosphere and become diluted. What remains is the air-borne part, which can be calculated by the nuclear bomb test studies (0.5% background). The early calculation around 1975 of cancer death rate of a reactor meltdown made by both the pro and con nuclear camps both suffered the above errors. But their near agreement with each other was taken as the *neutral truth* without knowing them to be both wrong, contradicted by the experimentally determined value of 0.5% of the background (an re-enactment of Rashomon). The radiation level generated by reactor explosion is so negligibly small that thirty years of hot debate on the low level radiation effect turned out to be much ado for nothing.

The beneficial effect of radiation is not surprising and can actually be anticipated by the folklore of Colorado and Chekoslovakia which deem that the uranium mine tunnels can impart a beneficial health effect, and also that of Brazilians who go to the beach (high uranium in sand) for health benefits. Radiation from nuclear waste may be used to make simulated uranium mine tunnels to provide health services. Great trash can than be turned into great treasure.

Preliminary work has been reported in invited papers to the American Physical Society[15,16] and an paper to the Symposium *A Decade After Chernobyl.*[11]

References

1 United Nations, Department of Economic and Social Development, *Economic and Social Aspects of Population Aging in Kerala, India,* ST/ESA/SER/119, New York, 1992. [Microfiche 510. International Information Service (1993), #3080-M104]

2. Willard F. Libby, Bull. Am. Phys. Soc. II **2**, 206 (1957).

3. Statistical Abstract of the United States, Government Printing Office, Washington, D.C., 1951 to 1996, a total of 46 volumes inclusive.

4. B. L. Cohen, Int. Arch, Occup. Health, **66**, 71-75 (1994).

5. Rand McNally, *Commercial Atlas And Marketing Guide*, 128th Ed., 1997.

6. Statistical Abstract of the United States, 1994, Government Printing Office, Washington, D.C.

7. B. L. Cohen, Health Physics, **68**, 157-178 (1995).

8. Lugwig E. Feinendegen, Bull. Am. Phys. Soc, **42**, 1124 (1997).

9. S. Kondo, *Health Effect of Low Level Radiation*, Kinki University Press, Osaka, Japan, 1993.

10. The World Journal, New York, Dec. 12, 1996.

11. United Nations, European Union, and World Health Organization, Symposium *One Decade after Chernobyl*, April 8-12, 1996, Vienna.

12. Myron Pollycove, Bull. Am. Phys. Soc. **42**, 1124 (1997)

13. Sun Laiyan, Marine Fishery **4**, 152-154 (1991). (in Chinese with English abstract).

14, Philip H. Abelson, Science, **265**, 1507 (1994).

15 Peter Fong, Bull. Am. Phys. Soc. **40**, 2075 (1995),

16. Peter Fong, Bull. Am. Phys. Soc, **41**, 1646 (1996).

Low Level Radiation Enhances Life Expectancy from the Study of Leading Diseases of the U.S.

Peter Fong

Physics Department, Emory University

Atlanta, Georgia 30322

Abstract and Conclusion

With minor deviations, the mortality statistics of the United States over the past half century covering a continent reveal that the leading diseases causing death are all reduced by a natural factor independent of time and human activities but dependent on topography that can be identified with the natural background radiation. We thus conclude that the voluminous data and their graphic representation strongly support that moderate increase of radiation enhances life expectancy.

Introduction

The beneficial effect of low level nuclear radiation against cancer has been studied widely and reviewed.[1] In the previous studies[2,3] of cancer mortality rates in the United States[4] we compared the 8 mountain states, Arizona, Colorado, Idaho, Montana, Nevada, New Mexico, Utah, Wyoming, as a whole with the entire nation and found the ratio of their mortality rates to be nearly a constant over the past 4 decades, averaged to 0.752 with a root-mean-square deviation of 2.5% (see Figure 1 of the preceding paper). It means that the mountain states have a cancer death rate one-fourth lower than the national average.

Since the mountain states have a background radiation twice larger than that at the sea level, it is natural to associate the reduction of cancer rates to the higher radiation as we have learned by earlier observations in other fields. However, by this experiment alone other causes for the cancer reduction cannot be excluded.

To exclude other causes we compared the 8 mountain states among themselves[2] to find if there is any quantitative correlation of cancer mortality rates with their altitudes. We found indeed that cancer rates decrease with increasing altitudes nearly linearly (see Figure 2 of the preceding paper). Thus the cause of cancer reduction is a factor that changes with the altitude nearly linearly. This eliminates most speculations on the causes, such as weather, fresh air and all social economic conditions, which do not vary with altitude linearly. In fact there are only two factors that vary nearly linearly with altitude, that is, the background radiation and the atmosphere pressure (and any factors dependent on them such as the boiling point of water).

By the slight difference of the deviations from linearity with altitude, which is traced to the difference of the functional dependence of radiation and atmospheric pressure on altitude, we can nail down the cause of mortality variations to the background radiation (see the preceding paper). Thus the cancer mortality rates statistics can be used to determine the exactly law of cancer causation by the variation of background radiation.

In the process of the previous studies[2,3] it was found accidentally that the pattern of cancer reduction in mountain states is repeated in many other leading diseases of death, such as heart and cerebrovescular diseases with a comparable proportion of reduction.[4] This is completely unexpected because there is no obvious connection of radiation with these degenerative diseases which are largely due to wear and tear of the physiological process. A program is thus carried out to study the 8 leading diseases that cause natural death, the work on each one of the 8 is a repeat of the study of cancer done previously, based on the published mortality rates.[4]. The 8 leading causes of natural death include cancer, 6 degenerative and 1 communicative diseases. Any one of the 8 may cause death. Only when all 8 are reduced can the life expectancy be increased substantially. Preliminary results show that indeed all the involved diseases are reduced by increasing background radiation level in the mountain states. Thus radiation increases may seem to enhance life expectancy. The detailed statistical results will be reported here.

The result that radiation exposure may increase life expectancy is a great surprise because in all previously years nuclear radiation has been looked upon in the detrimental perspective. However, the new results are supported by the recent discovery[5] that Kerala, India is characterized by an increase of life expectancy of 10.7 years[6] correlated with an enhanced background radiation that is 20 times the normal value.

Results

In Figures 1 to 8 we present the results of the 8 leading diseases respectively that cause death in the descending order of their mortalities. They are: heart diseases, cancer, stroke, degenerative (non-cancer non-communicative) lung diseases, pneumonia/flu, diabetes, liver, and hardening of artery. In each of them we plot the mortality reduction ratio (mountain state mortality rate divided by the national average) as a

Heart Disease Ratio: Mountain States/ Entire United States

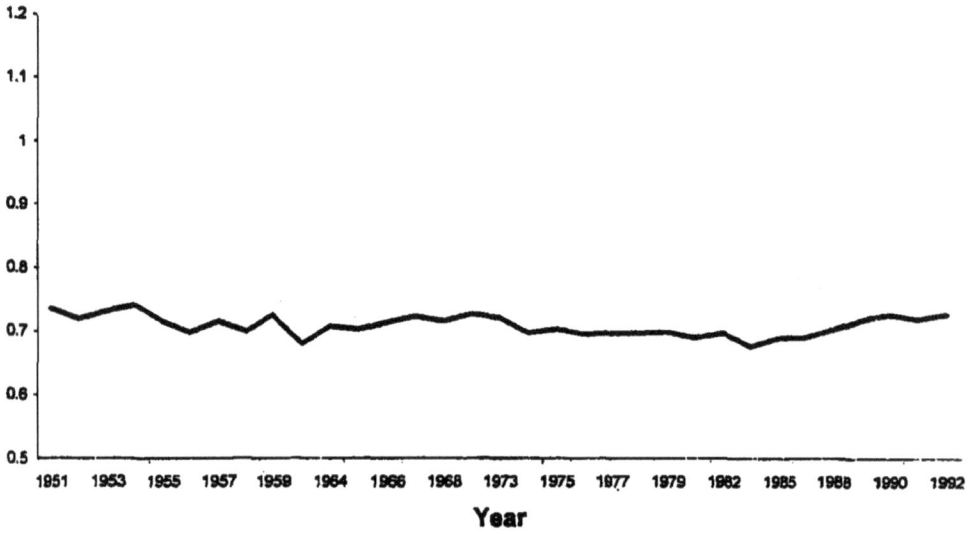

Figure 1

Cancer Ratio: Mountain States/ Entire United States

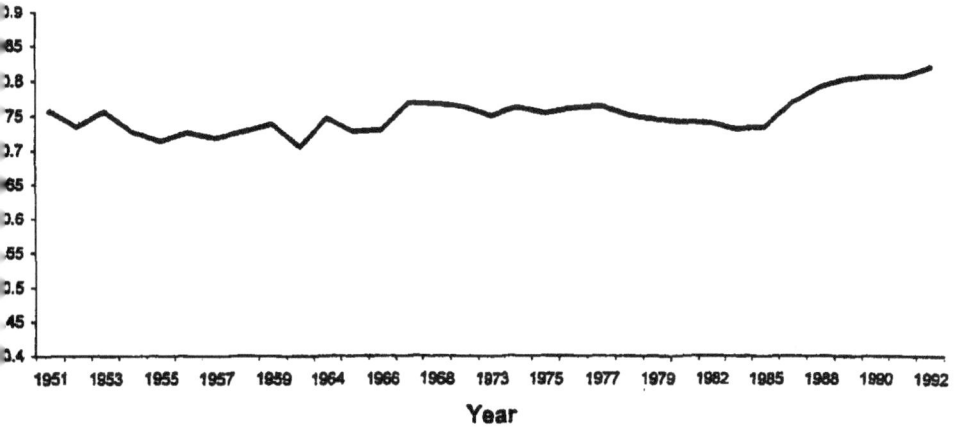

Figure 2

Cerebrovascular Disease Ratio: Mountain States/ Entire United States

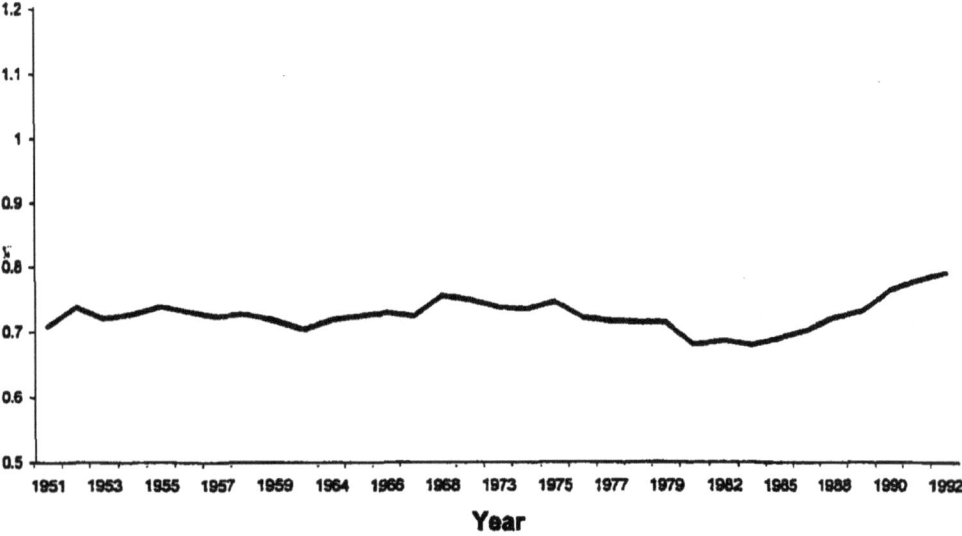

Figure 3

**Chronic Obstructive Pulmonary Disease Ratio: Mountain States/
Entire United States**

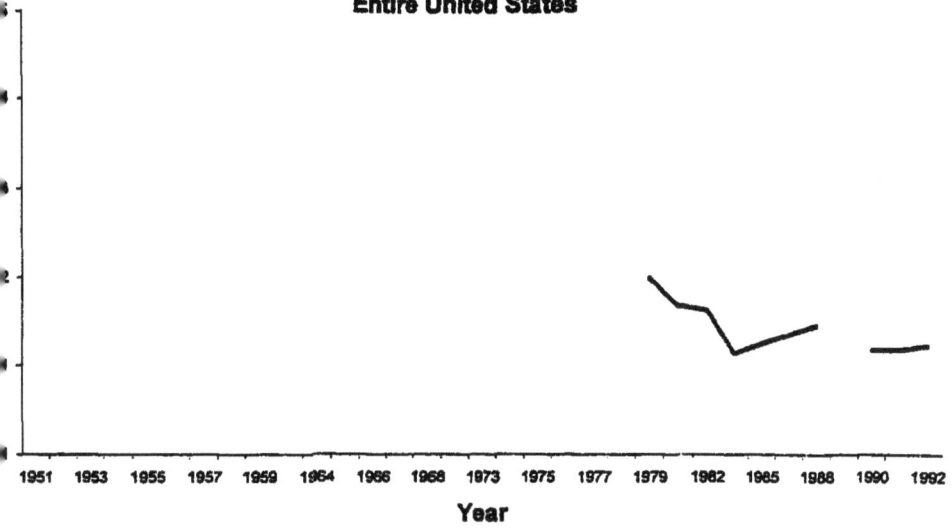

Figure 4

Pneumonia, Flu Disease Ratio: Mountain States/ Entire United States

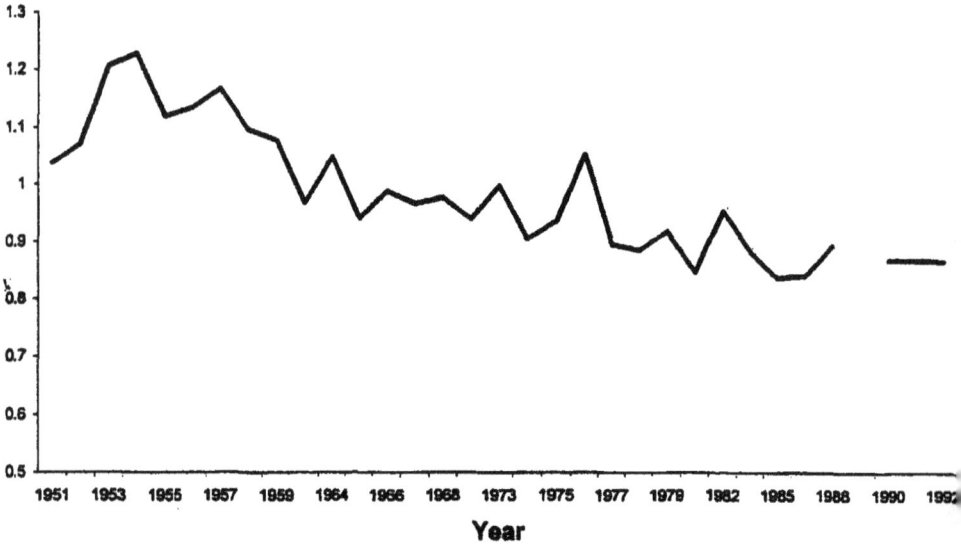

Figure 5

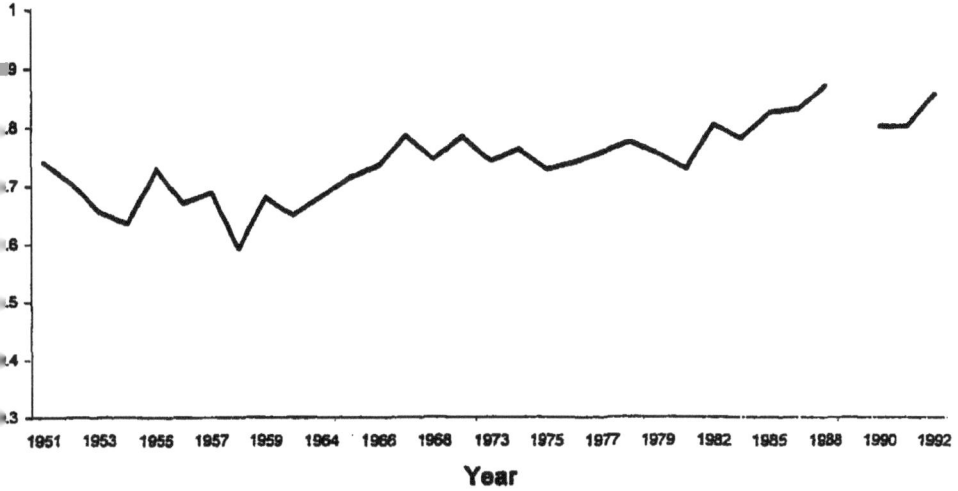

Diabetes Mellitus Disease Ratio: Mountain States/ Entire United States

Figure 6

Chronic Liver Disease and Cirrhosis Ratio: Mountain States/ Entire United States

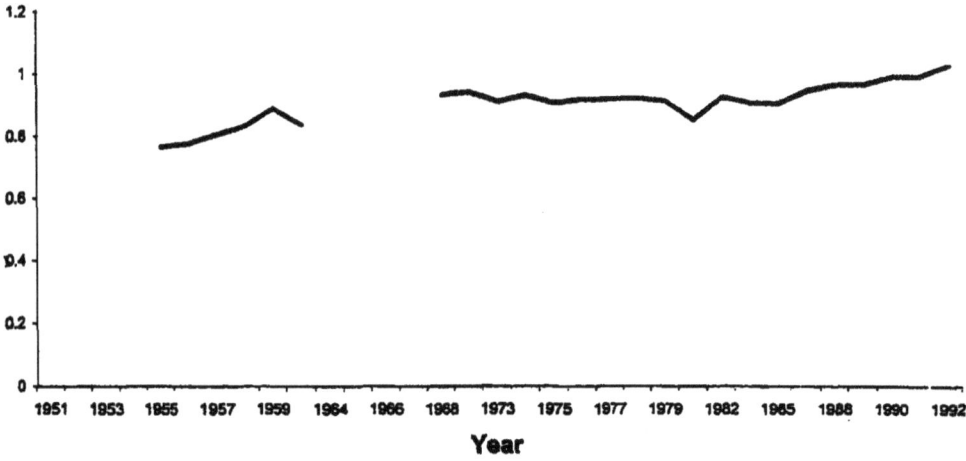

Figure 7

Atherosclerosis Disease Ratio: Mountain States/ Entire United States

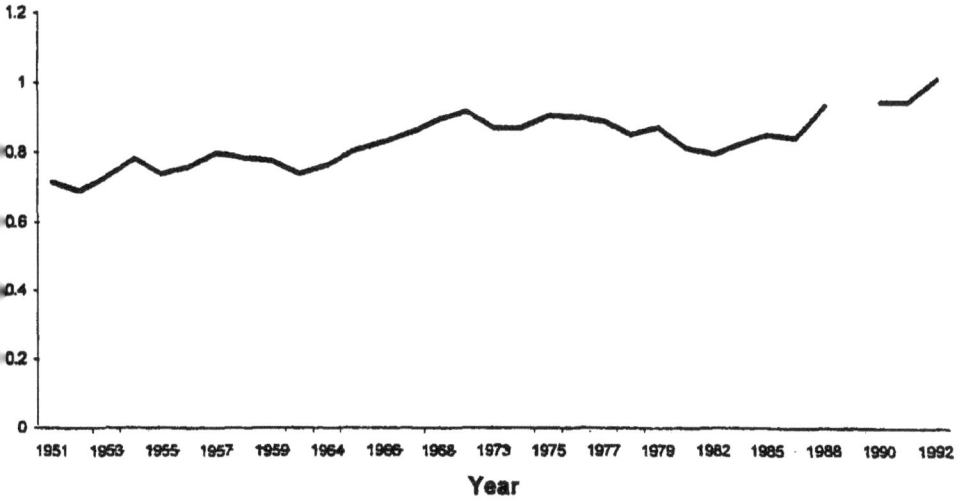

Figure 8

function of time (year) covering the past 4 decades. In other words we repeat the Figure 1 for cancer of the preceding paper for all the 8 leading diseases to death. Most of these 8 curves are flat with the ratio of reduction being nearly a constant for the entire period. This indicates that the reduction effect is due to a factor that does not change in time and independent of human activities, that is, a constant natural factor, which is identified with the background radiation just as we have done for cancer in the preceding paper.

Except the degenerative (non-cancer non-communicative) lung diseases, all seven show a decrease of the mortality rate in the mountain states like what is known with cancer (25% reduction). To be more specific the 7 diseases and the percentage of reduction of their mortality rates in the mountain states of US are listed as follows according to the order of mortality: heart disease, 27%; cancer, 25%; stroke, 23%; flu and pneumonia, 12%; diabetes, 20%; liver disease, 1%; hardening of artery, 5%. These ratios of reduction are nearly constant in the past 40 years. One exception of the constant rule is the communicative disease pneumonia/flu which came in random spurts of epidemics. A secular declining trend reflect the advance of medical services. We thus conclude most diseases are reduced by a constant factor.

The one exception to the decreasing trend of the 8 is the degenerative (non-cancer, non-communicative) lung diseases, which increased 11% for the mountain states. This may be explained as follows: The oxygen content in mountain states is about 16% lower than at sea level. Residents have to breathe harder to make up the 16% oxygen deficit. The wear and tear may cause this lung degenerative disease. Disregarding this kind of exception we may conclude in general that increase of background radiation may increase life expectancy.

This explanation also eliminates the conventional view that the beneficial health affect in mountain states is due to better air quality in the mountains because that would reduce the non-cancer non-communicative lung diseases. This also eliminates the possibility of less oxidant agents in the body from the 16% atmospheric deficit because the consumption of oxygen is not changed. Thus the longevity effect is entirely due to high background radiation, not due to any other factors of high altitude.

In connection with the previous study of the 8 mountain states separately to find the correlation of cancer mortality rates with altitudes (Figure 2 of the preceding paper), we repeat the same program for the 8 leading diseases and for all the years to make sure that the conclusion obtained has universal applicability. What we do is to break up each curve of one disease in Figure 1-8 into 8 curves for the 8 mountain states plotted together on the same graph but distinguished by different colors. The results are displayed in Figures 9-16.

Figure 2 of the preceding paper is just a cross section for a particular year of Figure 10 for cancer, which is analyzed in great detail. The same analysis applies to all other diseases and all other years and thus need not be repeated here. (In particular the exceptions of Utah and Montana have been explained previously as due to the abstinence and indulgence respectively of tobacco and alcohol; indeed these two exceptions stand out prominently as the top and bottom curves of all the graphs.)

The most outstanding feature of these figures is the parallelism of the curves of different colors (states) in each figure (disease) indicating the same altitude correlation of mortalities persisted all the time, and are

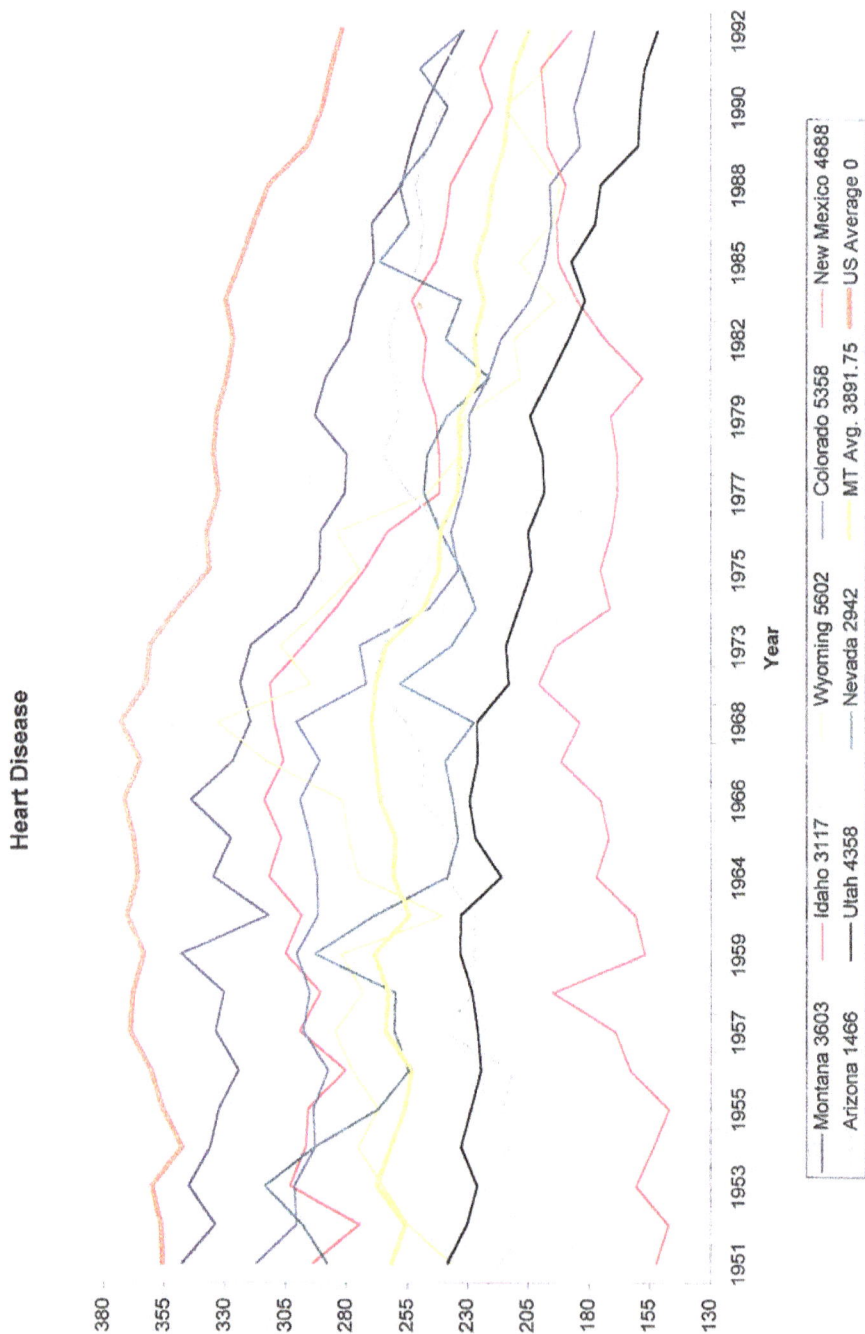

Figure 9

Cancer

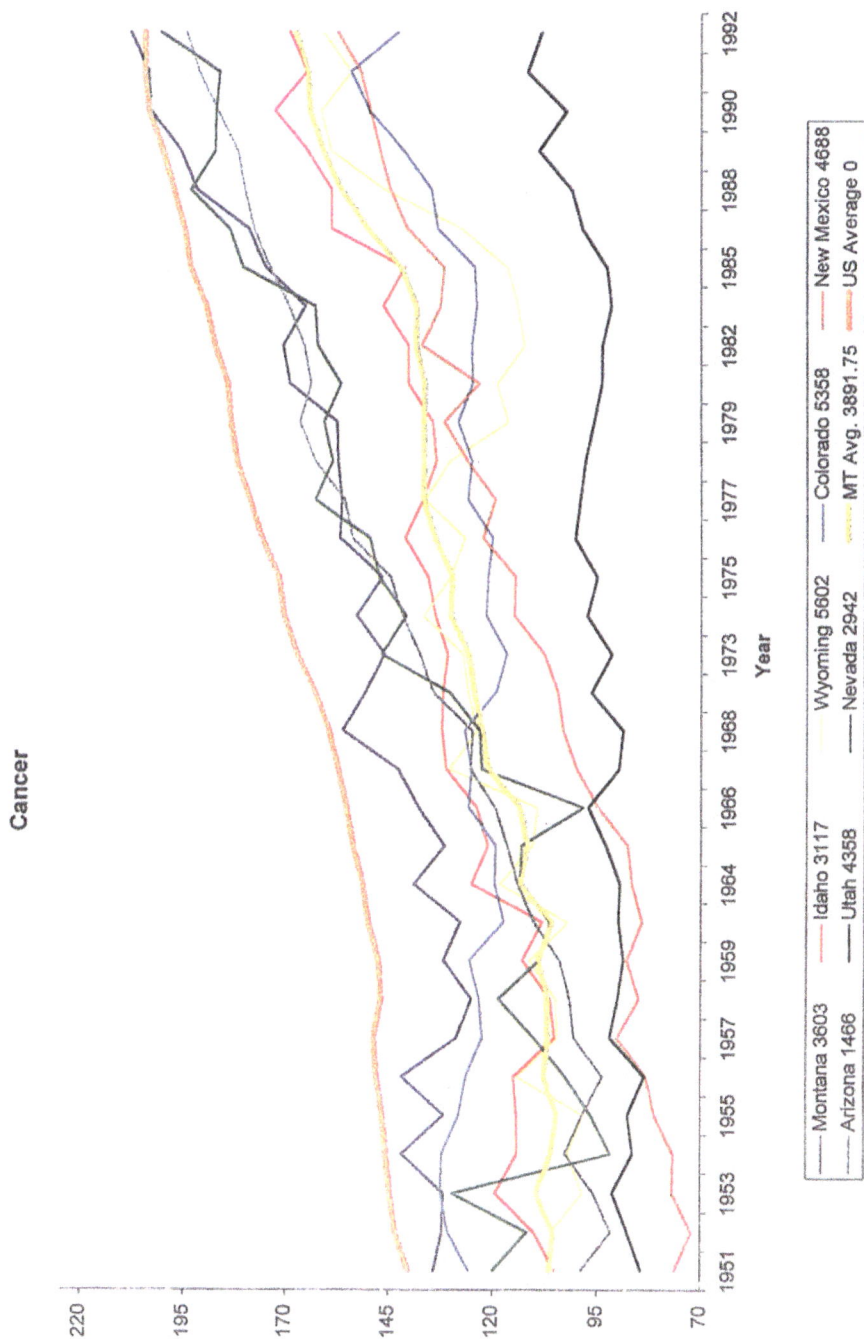

Montana 3603   Idaho 3117   Wyoming 5602   Colorado 5358   New Mexico 4688
Arizona 1466   Utah 4358   Nevada 2942   MT Avg. 3891.75   US Average 0

Figure 10

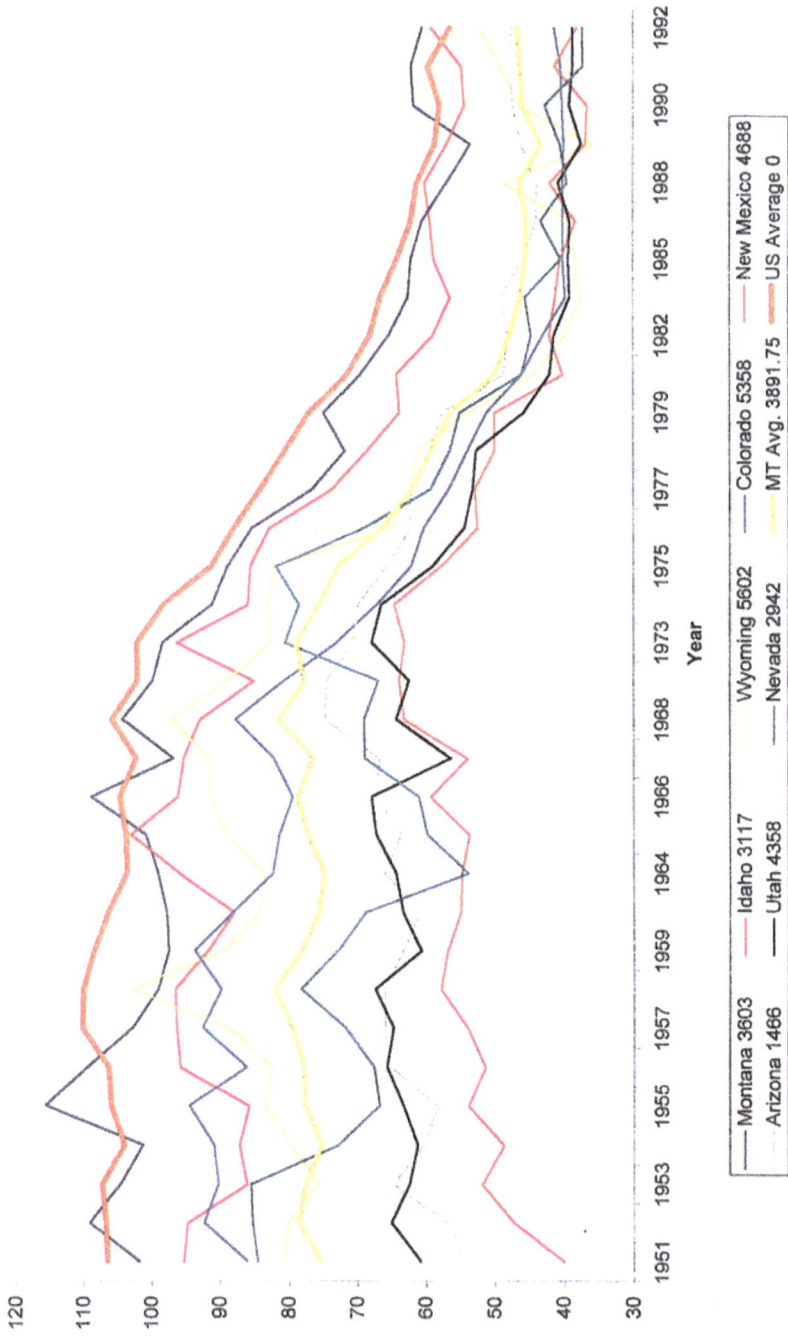

Figure 11

Pulmonary Disease

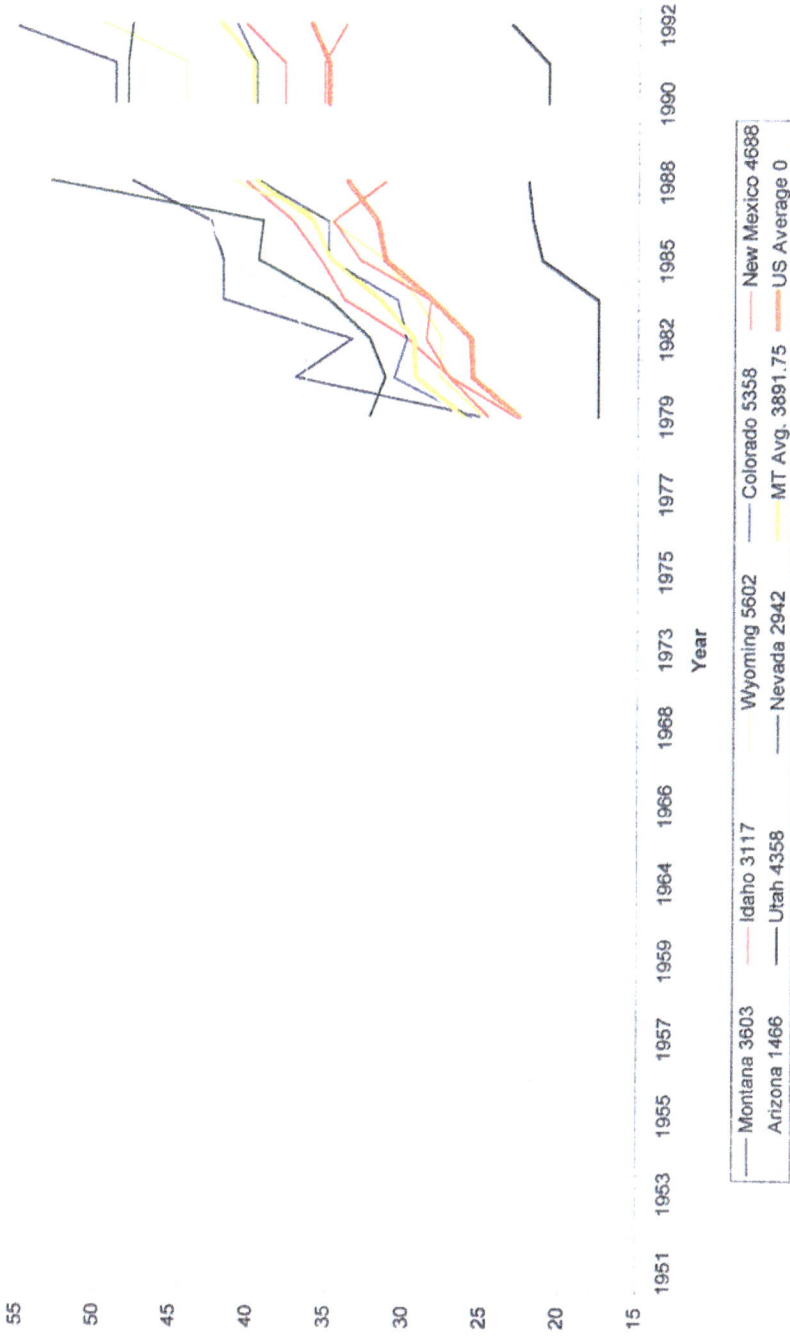

Figure 12

Legend:
- Montana 3603
- Arizona 1466
- Idaho 3117
- Utah 4358
- Wyoming 5602
- Nevada 2942
- Colorado 5358
- MT Avg. 3891.75
- New Mexico 4688
- US Average 0

Year

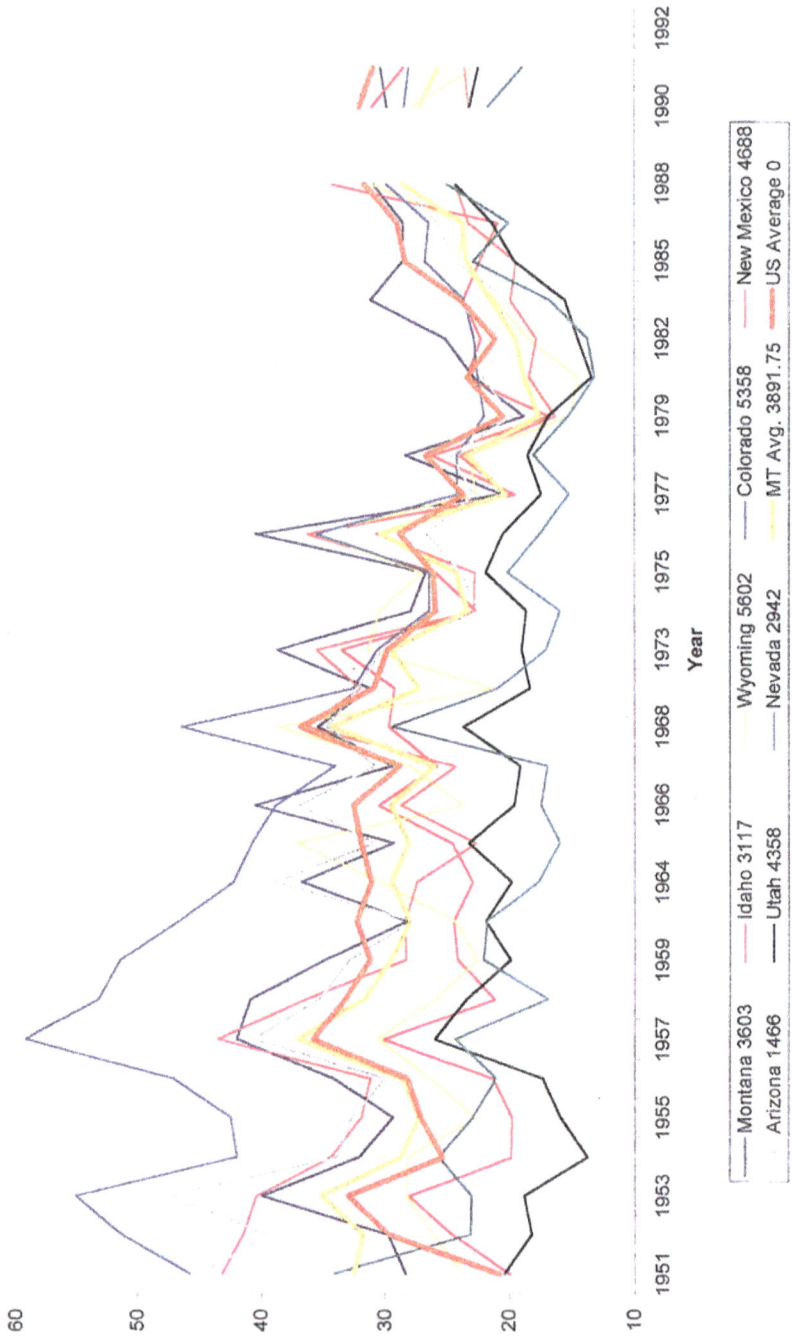

Figure 13

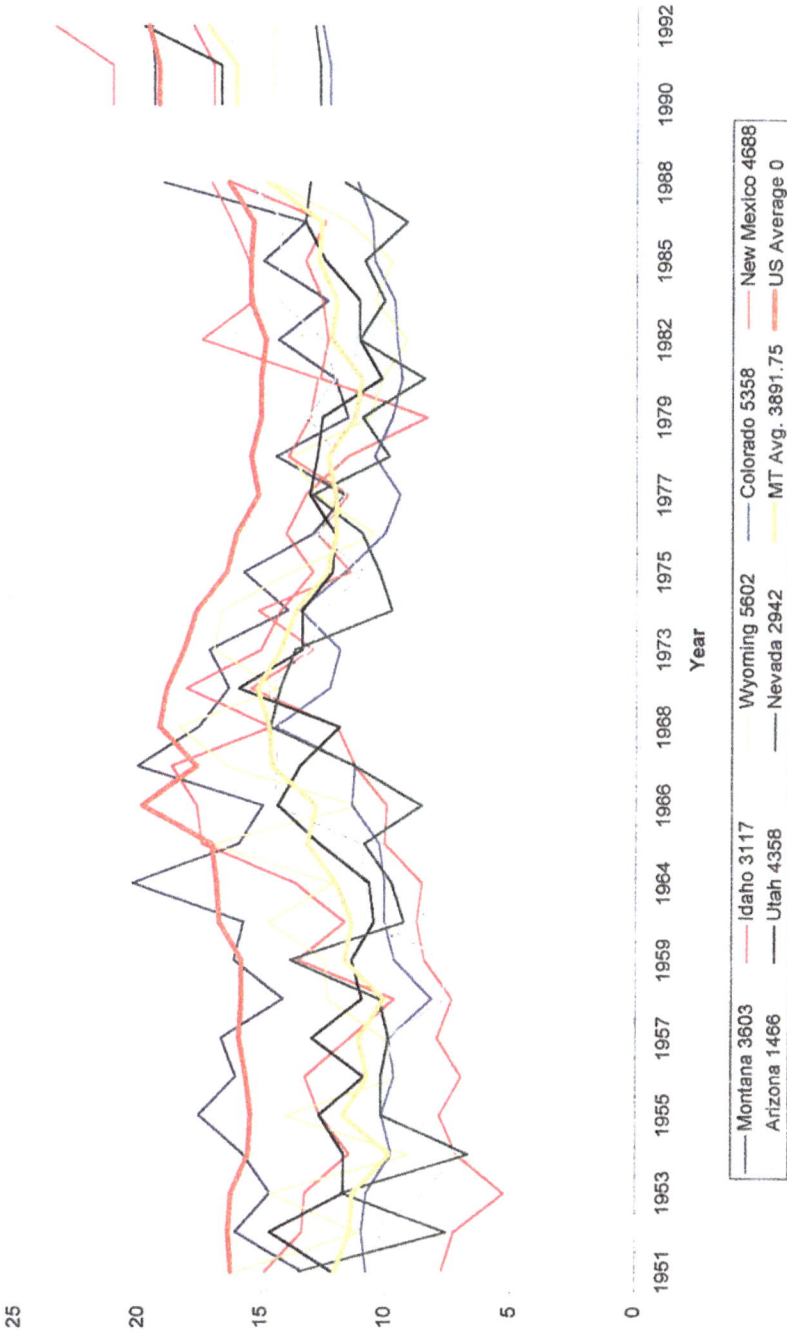

Figure 14

Diabetes

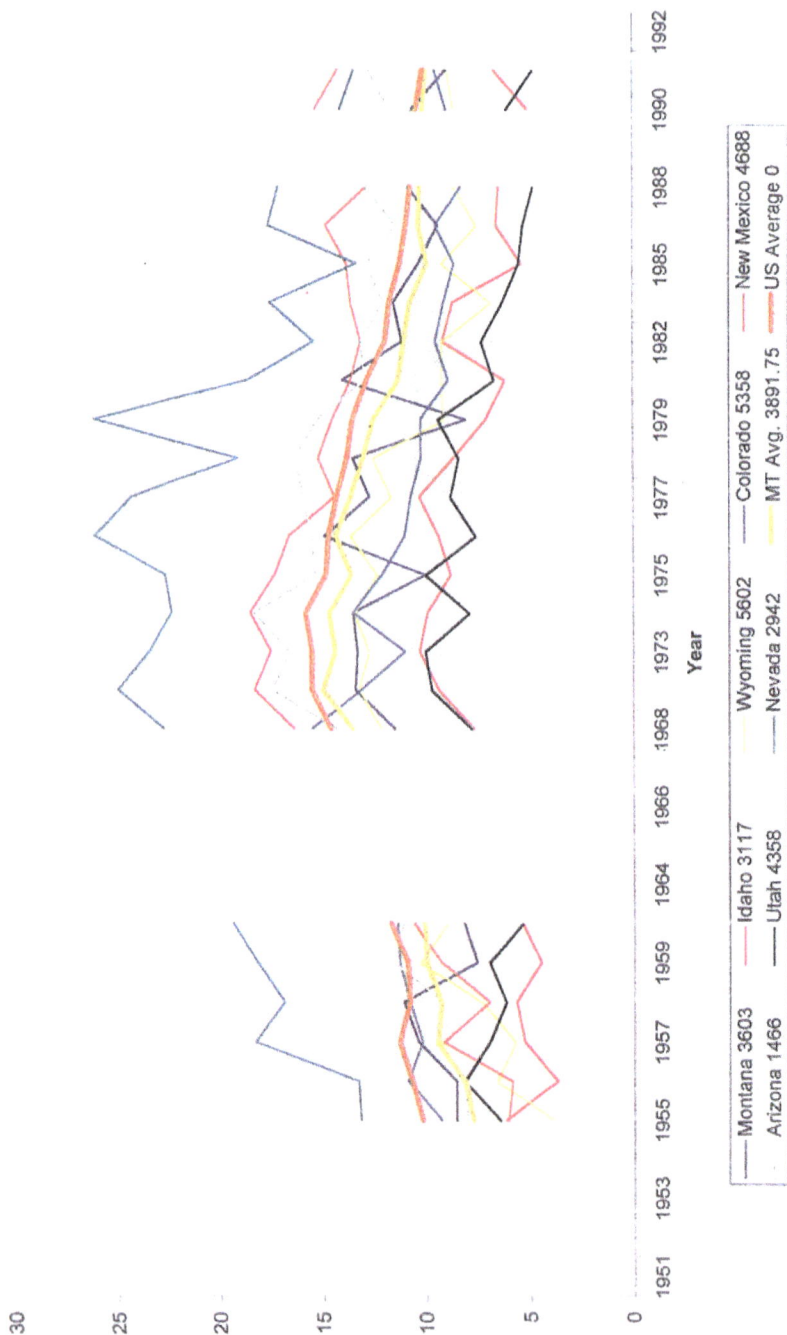

Figure 15

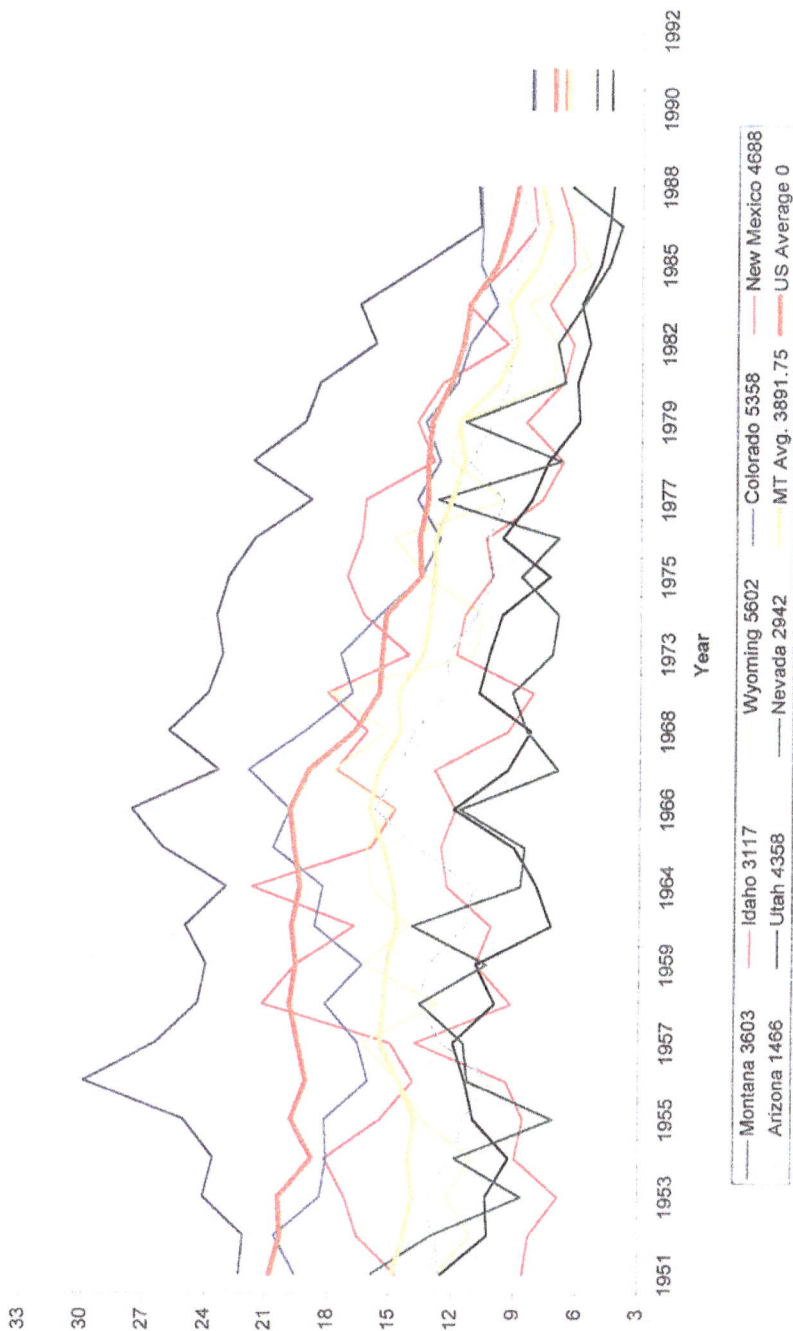

Atherosclerosis

| Montana 3603 | Idaho 3117 | Wyoming 5602 | Colorado 5358 | New Mexico 4688 |
| Arizona 1466 | Utah 4358 | Nevada 2942 | MT Avg. 3891.75 | US Average 0 |

Figure 16

generally the same for all the 7 diseases. Thus the general principle established in the preceding paper applies universally. In particular all 7 diseases are reduced by **increased** radiation and thus increased radiation enhances life expectancy. This is the most comprehensive **collection** of evidence to support the idea that radiation increases longevity.

The author wishes to thank Mr. David Bovi for preparing the graphic work of the paper, especially the colored graphs which present a large amount of information and bring out the correlations in the most efficient way.

References

1. B. L. Cohen, Int. Arch, Occup. Health, 66, 71-75 (1994).

2. Peter Fong, in A *Decade After* **Chernobyl,** a Symposium sponsored by UN, EU, WHO, Proceedings published by IAEA, Vienna, 1996

3. Peter Fong, Bull. **Am.** Phys. Soc, **41, 1646** (1996).

4. Statistical Abstracts of the United States, 40 volumes from 1957 to 1996. Government Printing Office, **Washington,** D.C.

5. Peter Fong, Post-deadline paper presented to American Physical Society, Washington, D.C., April **18-21, 1997**.

6. United Nations, Department of Economic and Social Development, *Economic and Social Aspects of Population Aging in Kerala, India.* **ST/ESA/SER/119,** New York, 1992 [Microfiche 510. International Information Service (1993), **#3080-M104**]

EUROPEAN COMMISSION · INTERNATIONAL ATOMIC ENERGY AGENCY · WORLD HEALTH ORGANIZATION

INTERNATIONAL CONFERENCE
ONE DECADE AFTER CHERNOBYL:
SUMMING UP THE CONSEQUENCES OF THE ACCIDENT

Austria Center Vienna, Austria, 8-12 April 1996

Paper number : IAEA-CN-63/.405.

Title: Reexamining Nuclear Energy Safety

Authors/affiliations: Peter Fong

Physics Department, Emory University, Atlanta, GA 30322, USA

Reexamining Nuclear Energy Safety

Peter Fong

Physics Department, Emory University

Atlanta, Georgia 30322

Abstract

Recent studies show that a low-dose carcinogen may *reduce* cancer risks. It will be shown quantitatively that radiation will do the same. Thus nuclear energy risks have been greatly over-estimated, which can now be corrected. We found the ratio of cancer mortality rate of the eight mountain states to the entire US is a *constant* 0.752 over the past 40 years. The reason for this cancer reduction thus cannot be any human or biological factors which would change over 40 years. It must be chemical or physical factors that is constant and pathogenic, i.e., the natural background radiation, which increases 100% in the mountain states. Thus we conclude a doubling of the background radiation will reduce cancer mortality rate 25%. This calculation is free from the criticism of epidemiologists from the viewpoint of ecological fallacy. The radiation released in the worst nuclear plant accident happens to be an amount that would double the background radiation. Instead of causing 120,000 cancer deaths in 30 years believed previously, it would save 3 million lives from all cancer deaths and therefore nuclear energy is safe. Actually even without the beneficial effects, 120,000 deaths in 30 years means 4000 deaths a year, a risk comparable to that of drowning (3967 deaths a year). Thus a "typical" reactor meltdown, when actually happened, is as risky or as safe as a backyard swimming pool. The grave concern on nuclear power safety in the past 25 years was not real, even without regard of its beneficial effect.

1. Introduction

Abelson's Editorial in *Science* of September 9, 1994 on the beneficial effects of low-dose carcinogen exposure[1] and the seven Letters to the Editor in response[2-8] shed light on a similar issue of cancer risks of low level radiation which determines the nuclear power plant safety with equally startling results. We found the nuclear plant even in the worst possible accident may not be disastrous but beneficial. It may save a large number of cancer deaths that routinely happen from all causes.

2. Low dose carcinogen may reduce cancer risks

The scientists mentioned above emphasized that the living systems actively respond to environmental assaults in damage repairs (amply known in bacteria long ago) and therefore the simple-minded linear extrapolation of radiation risks to low-doses without threshold as if dead material is not appropriate. Unexpected animal experiment results reported recently in a large volume show that a small dose of carcinogen, such as dioxin, can actually reduce, not increase as expected, all cancer risks.[1,2,3,4,5,7] Actually this should not be surprising since a weakened polio virus can activate the immune system against polio. Likewise the DNA repair mechanism may be stimulated by a small dose of carcinogen to reduce cancers. Thus low-level exposure may be beneficial and so may have a safety threshold, which changes risk assessment completely. The current practice of risk assessment on the basis of linear extrapolation to low doses without threshold is unnecessarily conservative and wasteful.

3. Low dose radiation may do the same

What the nuclear radiation does to DNA is not much different from that of a carcinogen. Therefore low-dose radiation exposure may likely stimulate the DNA repair mechanism to reduce all cancer risks like what dioxin has done. Organisms survived the struggle for existence on the earth surface must have developed DNA repair mechanisms adequate to deal with damages from natural background radiation, 100 millirem per year at sea level, lest they become extinct. It is thus reasonable to say that low-dose radiation may have a beneficial effect and a safety threshold of the order of the natural background radiation.

The enhancement of DNA repair is likely to be mediated by a negative feedback of the radiation assault by increasing the on-going DNA repair mechanism, as happened often in many automatic mechanisms carried out by enzymatic activities in molecular biology. Studies of cell mechanisms during DNA repair uncovers enzymatic feedback mechanisms to slow down the cell cycle in the S phase so that more time will be available for DNA synthesis and repair. In a similar way radiation assault on DNA may likely induce enhanced DNA repair in response.

Radiation has a characteristic quantum manifestation. One quantum of it from alpha, beta or gamma rays can cause DNA damage. This provides the often mentioned rationale that radiation effect may have no threshold. What is left out of the argument is the effect of DNA repair, particularly the low-level radiation induced, which would introduce a threshold. Once the radiation is diluted to the threshold, it becomes harmless.

In recent years there are experimental studies to show that previous exposure to low level radiation may reduce the cancer risks later in high dose radiation. In a recent review Cohen[9] has listed 12 references for reduction of chromosome aberrations for human and animal cells, and 4 references for the reduction of frequency of mutation of human and animal cells. Perhaps the most striking experiment is that previous exposure to low dose of X-rays can reduce the number of dominant lethal mutations of *drosophila*,[10] bringing out the vivid analogy of an immunization injection. Also Takai et al have found tumor remission even when the irradiated area does not cover the tumor.

Cohen also elaborated epidemiological evidence such as radiologists and their technicians, exposed to more radiation, actually have experienced lower cancer rates. The same applies to nuclear plant worker, nuclear test personnel, and residents near nuclear plants. Even the atomic bombs in Japan, while generating radiation diseases in high dose regions, have generated beneficial effects in low dose regions.

However, no quantitative law was given to specify the reduction of cancer rate with respect to increase of the low radiation level. This is the objective of this paper. The result is needed for a correct assessment of the risks of nuclear power plant safety.

4. Beneficial effect suggested by Libby four decades ago

As early as four decades ago the then Atomic Energy Commissioner Professor Willard F. Libby has reported to the American Physical Society that the cancer death rate in mile-high Denver with a doubled background radiation level of 200 millirem per year is actually smaller, not larger as expected, than Los Angeles with the sea level background radiation of 100 millirem per year.[11] It anticipated the recent dioxin results and suggested that low dose radiation has a beneficial effect in reducing cancer. But at that time this idea was so outlandish that it was ignored. Everybody thought radiation at any level is harmful. The smaller cancer rate of Denver was thought to be a statistical fluke, but was good enough to show that an increase of 100 millirem a year is harmless. In fact radiation safety standard was established on this basis. With the recent results on dioxin and so on, we may now pursue the idea earnestly. The updated 1994 data[12] of per 100,000 population cancer deaths annually for the US is 204.1. The 8 mountain states have the lowest rate among 10 regions of the nation at 164.9 and Colorado including Denver is the second lowest state at 154.2, verifying Libby's point and showing a substantial 25% reduction of cancer death rate.

To appreciate this large rate of reduction, one must realize that DNA damage and repair appear all the time frequently. In DNA replication there are thousands of base mistakes in one cell division, most of which are repaired. Thus the cell during replication is more like a repair garage than a manufacturing plant. Cancerous cells come and go, and cancer starts and stops all the time in all healthy persons. Only when the cancer cells reach a critical mass (for the prostate cancer it is 3 mm^3), then the repair mechanism can no longer contain its growth (the inner cancer cells are shielded from attack from the immune system by the outer cancer cells) and the condition develops into a cancer clinically.

It seems that cancer is less due to evil agents from outside with human beings as the innocent victims. Instead cancer is a part of human life all the time. The major sources of cancer are not nuclear radiation

(1% of all cancers) and toxic chemical (1%) but oxygen and food, the very elements that sustain life. Whether it develops into a deadly disease depends on humans themselves, i.e., on the repair mechanism of the immune system. In this respect carcinogens and radiation play a more subtle role than we realize before. They are deleterious in high dose (overwhelming repair) but beneficial in low dose (enhancing the repair mechanism). This changes the risk assessment entirely.

With such a natural bustling repair activity, it can be expected that a negative feedback of the assault on DNA may be developed to enhance the repair mechanism, which will have a high survival value in evolution. The beneficial effect of otherwise deadly agents can thus be understood naturally.

5. Study of 40 years' cancer statistics of the mountain states

The cancer mortality rate data of the mountain states and other regions of the US over the past 40 years[12] are collected and analyzed. Instead of comparing Denver with Los Angeles, we compared the 8 mountain states as a whole with the entire US and found the cancer deaths ratio of the two is almost a constant over the past 40 years, averaged to 0.752 with a root-mean-square deviation of 2.5%. It means the mountain states have a cancer death rate one-fourth lower than the national average all the time. (Actually the entire US population should be subtracted of the 4.8% of the 8 mountain states, with a negligible increase of 0.9% of the final result.)

It turned out that the choice to compare the 8 mountain states with the entire nation is a *tour de force* that leads to a quantitative result that will have wide applicability. If we compare Denver with Los Angeles, or

Colorado with California, we will not obtain a ratio that is a constant over 40 years but obtain a variable result influenced by demographic factors. It will have only anecdotal value (unless corrected of human related factors), but not as a quantitative law that has universal applicability. By comparing the 8 mountain states as a whole with the entire US, these variable factors are either averaged out (such as the peculiarities of Utah and Wyoming), or divided out in taking the ratio (such as the changing of smoking habit and the economic development, which definitely increases cancer death rate), resulting in an unencumbered, *invariant* law relating the cancer rate to the altitude alone, which can be used to study other problems.

6. Narrow down the cause from altitude to background radiation

In many studies altitude is identified as the proxy for background radiation level, which increases with altitude because of reduced atmospheric absorption of incoming cosmic rays. While all radiation effects appear through the altitude, the converse is not necessarily true. Thus we must first eliminate other altitude effects than radiation that may affect cancer mortality rate.

In Denver the background radiation increases 100%, atmospheric pressure reduces 16%, boiling point of water lowers accordingly, and other changes are negligible. Most features of high altitudes such as cold winter and high insolation can be found in other regions that do not have a reduced cancer rate (Wyoming with the highest elevation has large skin cancer rate because of high solar UV but has the lowest total cancer rate of the nation except Utah).

A major relevant concern is oxygen as an oxidant. Since Denver has an oxygen content 16% lower than that at sea level, the reduction of oxidant activity might reduce cancer. However, oxygen content in the lung is not determined by the air but by biological regulatory mechanisms. US population in the mountain states are not native Indians but mostly White immigrants, whose genetic structure is geared to sea level atmospheric pressure. When living in mountain states, they simply breathe a little harder (leading to higher non-cancer lung diseases than the other 9 regions) or slow down a little bit to made up the natural need of oxygen at sea level. Thus oxygen content in the lung and in the body system is an invariant, not a relevant factor of consideration of cancer mortality. Furthermore, if it were, it would affect the degenerative diseases, the aging process, hardening of artery, diabetes, Parkinson's disease, Alzheimer's disease, and so on, of which there is no systematic clinical evidence.

7. Epidemiologists' objections and ecological fallacy

The 25% reduction of cancer rate at mountain state altitude is a significant empirical finding. But additional considerations are needed to formulate a law of nature that can be applied to study other problems in general. Statistical studies such as this have been criticized by epidemiologists based on the notion of ecological fallacy from the early work of Hickey et al[13] on cancer statistics to the recent work of Cohen[14] on the study of radon-cancer correlation. Cancer mortality rate is a complex quantity depending on dozens of factors. Ideally it should be studied theoretically and experimentally as a function of dozens of variables. But statistics are never collected with such minute detail and are usually given by averaging over the uninterested variables to show the average result of a single variable, such as radiation level. But the average of the causes does not necessarily correspond to the average effect, as is often the case in statistics. Thus the established correlation may turn out to be an ecological fallacy.

All previous works are plagued by criticisms based on this consideration. Cohen[14] devoted 12 of 14 pages of his paper to eliminate the influence of all other confounding factors including 54 social and economic variables to establish his correlation of cancer mortality with radon level. Without a viable quantitative law because of the ecological fallacy, many practical problems of radiation safety cannot be solved with maximum efficiency, but have to rely on the principle of maximum assurance of safety, which is often ridiculously inefficient, wasting hundreds of billions of dollars.[1]

We try to find a way to extricate ourselves from the ecological quagmire. One approach, though not the most general and dependent on lucky circumstances occasionally, but often applicable to most physical science problems, is to sort out all variables according to their orders of magnitude and then treat them separately one by one in a successive approximation in the descending order of the magnitudes. When dealing with the leading variable, all other smaller factors can be neglected in the approximation to simplify the problem. This process can be repeated until all variables are accounted for. When the magnitudes are separated widely, the approximation is excellent and will give a viable solution to solve practical problems.

To explain it by the example of the solar system. The leading factor of all causes is the gravity of the sun. The next factor, which is much smaller, is the gravity forces of the planets. An even smaller factor is the residual gravity effect, such as the oblateness of the sun and the general relativity effects. Newton took the first approximation by neglecting the second and third order small factors and discovered the simple inverse-square law of the gravity force. It explained the Kepler's laws (elliptical orbits of planets and so on) and launched the scientific revolution, resulting in the technological development leading to the modern society we are in today.

But Newton did not explain the advance of perihelion of the planets, which was once regarded as the final reserve of God's free will without Newton's control. But Laplace explained that by the very Newton law of gravity but applied to the planets, the second order small factor, with astonishing success, reducing the solar system to a completely mechanical system controlled exclusively by Newton's law. Still, there is an unsolved problem, the advance of perihelion of Mercury, which can now be explained by the third order small factors. But this has no practical significance; the dominant position of Newton's law is established without doubt and any practical problems, such as space travel, will be and have been solved on that basis.

If Newton were to start with the advance of perihelion of Mercury, he would accomplish nothing. There would be no Newton's law, no scientific revolution and no modern technological society; we would still live under medieval theology. The success of Newtonianism is not entirely a matter of logic. Newton's law is established on crude experimental data but Laplace applied it to much finer experimental data of advance of perihelion, which is not logically justified. The success of Newtonianism is based on a natural philosophy that a simple law can explain almost all observed data; the validity of that law is then established by voluminous experiments taken together collectively.

The issue of low level radiation will be unraveled in this light. What the epidemiologists want is like to ask Newton to start with the advance of perihelion of Mercury to establish a logical theory. It is impossible and also unnecessary (and will throw us back to middle ages). Indeed the current safety policy on low level radiation is truly a medieval one, unnecessarily wasting resources based on "theology."

We can extricate ourselves from the complexity of dozens of confounding factors by first ascertaining their orders of magnitude and then do successive approximations. We find the most influencing factor on cancer mortality change is the radiation level (25%), the next is cigarette smoking (16%), all other factors are third order small (about 1%). Cancer is very difficult to change (that is why it is so hard to cure). Air pollutants CO, SO_2, HC's, NO_x are all not carcinogens.

Thus we may start with a first order approximation by neglecting all factors except the leading one, the radiation level. We use the "crude" data of cancer mortality of the mountain states to establish a dominant first order relation that a doubling of the radiation level reduces the cancer mortality by one-fourth (analogous to Newton's study of earth's orbit). No matter what we learn eventually on cancer mortality, this is a result established by 40 years' cancer statistics, which cannot be ignored and must be explained. (It may be mentioned in passing that in taking the ratio of the mountain states over the entire US, we have the largest population samples for the numerator and denominator and gain the advantage of statistical accuracy in carrying out the crucial step of our first approximation. As the samples are the largest possible, no other statistical study would have a better accuracy.)

From this crude result as the first step, we theorize a more universal law that radiation level affects cancer mortality rate *linearly* with the linear coefficient given by the first order result (analogous to Newton's study of other planetary orbits). This is a simple reasonable law based on our knowledge of molecular biology and evolution. It would be important and useful if its assertions can be verified empirically, which we strive to prove.

First we want to establish the universality of the proposed law for all kinds of radiation (analogous to Newton's study of the moon orbit). The above mentioned empirical relation is established for the change of geographical background radiation. To establish it for the other types of radiation we may take advantage of Cohen's study of radon-cancer correlation. The radiation concerned is the indoor radon level. From his correlation curve we can read out, for an increase of radon radiation level of 100 millirem per year, the decrease of lung cancer rate and then convert it to the decrease of all cancer rate. The result is a decrease of 26%, which agrees with our value of geographical background radiation effect of 25%. Thus the law appears to apply to different kinds of radiation.

For the proposed law to be universal the 8 mountain states, having slightly different altitudes and thus different background radiation, should have proportionally different cancer mortality rates, which can be used to verify the linear relationship assumed in the proposed law. Preliminary study of the cancer mortality rates of the 8 mountain states separately indeed displays a linear relation with their altitudes but there is a marked exception in the case of Utah with much lower value of mortality rate than indicated by the smooth curve.

If this exception could not be explained away, our "natural philosophy" would appear to be wrong and should be abandoned. On the other hand it is likely that other second order factors may have entered to change the simple relation determined from the altitudes. We search through possible influencing factors, the next large one being cigarette smoking. Indeed, Utah is populated largely by non-smoking Mormons, and this explains Utah's exceptionally small cancer mortality rate very nicely (analogous to the explanation of the Mercury abnormality by the theory of general relativity).

Smoking variation is a small factor in general. Except a small difference between the city and country, smoking habit is generally the same among the population and can be averaged out. The only outstanding exception is Utah because of the ethnic peculiarity of the Mormon population. Thus all major factors are accounted for and together the proposed law is verified without major exception. Whatever remaining small factors should not affect the general law so established in a significant way. Incidentally all criticisms on previous studies based on ecological fallacy single out smoking as the confounding factor that messed up the statistical conclusion.

Thus we may conclude a general law that cancer mortality rate decreases linearly with increasing radiation level with a doubling of the radiation level related to a reduction of cancer mortality rate by one-fourth. This law can now be applied to study practical problems, such as the risk of a nuclear power plant meltdown which changes the magnitude of the background radiation in a range the law is known to be valid.

Incidentally in the study of the separate problem of smoking hazards, the problem of ecological fallacy is avoided when we lump all smokers together and take the ratio of mortalities of smokers and non-smokers, smokers being defined as those who smoke one pack or more a day. Actually one-pack smokers, two-pack smokers,..., have different mortality rates. If the pack distribution is wide, we cannot take a single ratio and will be in the quagmire of ecological fallacy. It so happens that the distribution is narrow and can be replaced by a single group (a delta function). A similar lucky circumstance exists in our study. The altitude distribution of the 48 states is such that the 8 Rocky Mountain states stand out as a separate group over a flat background, and can be treated in a way like lumping all smokers together. If the altitude distribution were continuous and wide, we would not be able to carry out a clear cut, solidly established first approximation to start the successive approximation method and the problem of ecological fallacy would be inextricable from the beginning. Indeed this is exactly the problem confronted by Hickey et al.[13]

8. Application to nuclear safety

The linear law we have established that a doubling of the background radiation will reduce the cancer death rate by one-fourth can now be applied to study nuclear energy safety issues. The increased radiation is now coming from the power plant meltdown.

Consider a nuclear power plant accident, such as that in Chernobyl. In the worst possible case, it will be a maximum meltdown with explosion and adverse wind dumping radioactivity in a large population center nearby, an event so rare that it would happen only once in one billion reactor-years. The *acute* effect, such as the death of 31 firemen at Chernobyl, is limited. In fact, Western reactors use water, not graphite as used at Chernobyl, as moderator, which will not catch fire and thus there will be no firemen at risk for high dose exposure. With proper evacuation the acute deaths can be reduced to zero. In fact, a reactor without an evacuation plan, such as the Shoreham plant in Long Island, is not permitted to operate and indeed Shoreham plant has been abandoned.

Most nuclear concerns concentrate on the *chronic* effect of low level radiation which would last for 30 years. It was thought the worst possible accident would cause cancer deaths over a 30 year period (almost exclusively by low level radiation) of 120,000 persons according to pessimist Henry Kendall or 50,000 according to optimist Norman Rasmussen. Both estimates are based on the method of linear extrapolation to low doses without threshold, which was regarded as the prudent measure in the absence of any definite experimental information but now we have definite experimental results to the contrary.

According to this method of calculation (and with BEIR Report data) the natural background radiation of 100 millirem per year would generate 130,000 cancer deaths in 30 years. Thus the worst Kendall estimate of the worst possible nuclear accident is equivalent to doubling the background radiation from 100 to 200 millirem per year. In other words, to bring the US radiation level to that of mile-high Denver. Now we know from the established law that radiation level would reduce all cancer rate by one-fourth, saving 100,000 cancer deaths out of an annual cancer fatality of 400,000 for the US or saving 3 million lives in 30 years. Thus instead of 120,000 cancer deaths in 30 years, the worst possible nuclear accident would save 3 million cancer deaths from all cancer sources in the same 30 year period and therefore would be extremely beneficial instead of harmful.

Actually the worst estimate of 120,000 deaths in 30 years means 4000 deaths a year, a risk comparable to that of accidental drowning (3967 deaths a year) or 1% of the risk of dying by cigarette smoking (360,000 deaths a year). It is a small effect. A "typical" meltdown, when actually taking place, would be as risky or as safe as a backyard swimming pool. It was the inflating of the number to 120,000 by the sly commercial gimmick of counting a long time of 30 years (which is often done in many similar cases) that makes a small effect appear as a major catastrophe. The propaganda effect was tremendous; the whole world turned against nuclear power. Even the nuclear industry itself is not excepted. It belabored to show the *probability* and thus the *expectation* of the nuclear accident to be extremely small, which cannot stop the mass hysteria just as the extremely small gambling odds cannot stop the mass hysteria of buying lottery tickets. But it never bothered to divide 120,000 deaths by 30 to show the effect of an *actual* meltdown to be very small, comparable to that of a backyard swimming pool. All massive calculations of Kendall and Rasmuson are much ado for nothing. This is a historical episode that is at once tragic and comic. If there is still a last doubt on nuclear safety, it can be removed by the beneficial effects of the low level radiation.

Thus nuclear power plant appears to be quite safe under all circumstances. All nuclear issues will undergo an about-face. This includes the stumbling difficulty of the problem of nuclear waster disposal, which was thought to be almost insurmountable, but now it is no longer intractable because the radiation is low level and may be beneficial. All low level radiation risks, including still birth and genetic defects due to DNA damage now fall on their own weight.

9. Cancer rate after Chernobyl and conclusions

The only additional relevant information on the issue is the cancer rate in the affected area after the Chernobyl accident. The earlier IAEA-

WHO study (no published report) found no increase of cancer due to radiation (cancer rate observed after Chernobyl in 1986 increases at the same rate as before since 1976). There was no increase of leukemia, which was the main cancer concern in nuclear accident due to the fission product Sr-90. The only thing certain that has been learned recently is a total of 171 children getting thyroid cancer (which may lead to 13 deaths) out of a population of 50 million in the affected area from 1986 to 1993 due to the fission product I-131. Thus the cancer risk is far less than the thousands of cancer deaths alleged by the news media. (Note added after Conference One Decade After Chernobyl: The Conference conclusion is that there is no increase of cancer after the Chernobyl accident.)

Abelson ended his Editorial[1] by saying that "The public has been needlessly frightened and deceived, and hundreds of billions of dollars wasted." The same can be said of the supposed nuclear risks. Assured by the new knowledge on the beneficial effect of radiation, nuclear energy can be utilized safely and the prospect of future world will be full of hope.

The author wishes to thank Bernard Cohen for discussions of the radon issue, Andrew Chen for discussion of his work, Fred Metler for the Chernobyl study, Hans Bethe and Alvin Weinberg for general discussions.

References

1. Philip H. Abelson, Editorial, Science **265**, 1507 (1994).

2. A. M. Monro, Science **266**, 1141 (1994).

3. C. J. Portier, G. W. Lucier,and L. Edler, Science **266**, 1141 (1994).

4. R. P.Zendzian, Science **266**, 1142 (1994).

5. M. Fox and M. Messelson, Science **266**, 1143 (1994).

6. B. S. Strauss, Science **266**, 1143 (1994).

7. R. C. von Borstel, Science **266**, 1144 (1994).

8. O. Ennemoser, Science **266**, 1145 (1994).

9. B. L. Cohen, Int. Arch. Occup. Health 66, 71-75 (1994).

10. H. Fritz-Niggli and C. Schaeppe-Bueche, Int. J. Rad. Biol. **59**, 175 (1991).

11. Willard F. Libby, Bull. Am. Phys. Soc. II 2, 206 (1957).

12 Statistical Abstract of the US, Government Printing Office,1955-1994.

13. Richard Hickey et al, Health Physics, **40**, 625-641 (1981).

14 B. L. Cohen, Health Physics, **68**, 157-178 (1995).

Index